The Role of Biotechnology in Countering BTW Agents

T0184106

NATO Science Series

A Series presenting the results of activities sponsored by the NATO Science Committee. The Series is published by IOS Press and Kluwer Academic Publishers, in conjunction with the NATO Scientific Affairs Division.

A. Life Sciences	IOS Press
B. Physics	Kluwer Academic Publishers
C. Mathematical and Physical Sciences	Kluwer Academic Publishers
D. Behavioural and Social Sciences	Kluwer Academic Publishers
E. Applied Sciences	Kluwer Academic Publishers
F. Computer and Systems Sciences	IOS Press

1. Disarmament Technologies	Kluwer Academic Publishers
2. Environmental Security	Kluwer Academic Publishers
3. High Technology	Kluwer Academic Publishers
4. Science and Technology Policy	IOS Press
5. Computer Networking	IOS Press

NATO-PCO-DATABASE

The NATO Science Series continues the series of books published formerly in the NATO ASI Series. An electronic index to the NATO ASI Series provides full bibliographical references (with keywords and/or abstracts) to more than 50000 contributions from internatonal scientists published in all sections of the NATO ASI Series.

Access to the NATO-PCO-DATA BASE is possible via CD-ROM "NATO-PCO-DATA BASE" with user-friendly retrieval software in English, French and German (WTV GmbH and DATAWARE Technologies Inc. 1989).

The CD-ROM of the NATO ASI Series can be ordered from: PCO, Overijse, Belgium

Series 1: Disarmament Technologies – Vol. 34

The Role of Biotechnology in Countering BTW Agents

edited by

Alexander Kelle

Peace Research Institute Frankfurt,
Frankfurt am Main, Germany

Malcolm R. Dando

Department of Peace Studies,
University of Bradford,
Bradford, United Kingdom

and

Kathryn Nixdorff

Institute for Microbiology and Genetics,
Technical University Darmstadt,
Darmstadt, Germany

Kluwer Academic Publishers

Dordrecht / Boston / London

Published in cooperation with NATO Scientific Affairs Division

Proceedings of the NATO Advanced Research Workshop on
The Role of Biotechnology in Countering BTW Agents
Prague, Czech Republic
21–23 October 1998

A C.I.P. Catalogue record for this book is available from the Library of Congress.

ISBN 0-7923-6905-X (HB)
ISBN 0-7923-6906-8 (PB)

Published by Kluwer Academic Publishers,
P.O. Box 17, 3300 AA Dordrecht, The Netherlands.

Sold and distributed in North, Central and South America
by Kluwer Academic Publishers,
101 Philip Drive, Norwell, MA 02061, U.S.A.

In all other countries, sold and distributed
by Kluwer Academic Publishers,
P.O. Box 322, 3300 AH Dordrecht, The Netherlands.

Printed on acid-free paper

All Rights Reserved
© 2001 Kluwer Academic Publishers
No part of the material protected by this copyright notice may be reproduced or utilized in
any form or by any means, electronic or mechanical,including photocopying, recording or by
any information storage and retrieval system, without written permission from the copyright
owner.

Printed in the Netherlands.

CONTENTS

Acknowledgements

Putting together an edited volume like the present one, one usually accumulates a number of debts to various people and institutions. It is with great pleasure that these debts are now made public. First of all, I would like to thank the NATO Science Committee for supporting the proposal that led to the Advanced Research Workshop whose presentations form the basis of the book chapters that follow. In addition, Nancy T. Schulte of the NATO staff has been very helpful throughout the project.

Special thanks are due to ARW co-director Jiri Sedivy of the Czech Institute for International Relations in Prague, who selected a great conference location and whose staff did practically all the 'ground work' during the workshop. At the Peace Research Institute Frankfurt, Harald Müller not only made project funds available for this endeavor but also allowed the writer of these acknowledgements to spend more time on the completion of this volume than was originally anticipated.

Without the untiring efforts of Malcolm R. Dando and Kathryn Nixdorff, my two co-editors, completion of the book would have taken even longer. Both of them also contributed considerably to the substance of individual chapters and the coherence of the book as a whole. Many thanks for this as well as for complying with deadlines that sometimes must have appeared arbitrary.

Gudrun Weidner deserves special credit for proof-reading and correcting several parts of the book, for providing the layout and – last, but certainly not least – for compiling the index. Thanks also to Gerard Holden who through his language editing improved the clarity of many passages of the book.

Lastly, a thank you goes to Annelies Kersbergen at Kluwer Academic Publishers for her cooperative spirit and – not to be underestimated – her patience.

Alexander Kelle   Frankfurt am Main, December 2000

BIOTECHNOLOGY IN COUNTERING BIOLOGICAL AND TOXIN WEAPONS: AN OVERVIEW

ALEXANDER KELLE
MALCOLM R. DANDO
KATHRYN NIXDORFF
GRAHAM S. PEARSON

There has long been recognition of the increased danger that can result from the application of biotechnology to biological and toxin weapons – whether through the greater ease of production or the development of genetically modified agents. This book is based on the NATO Advanced Research Workshop held in Prague in October 1998, which examined the significant potential that biotechnology has in countering biological and toxin weapons (BTW). For convenience, the role of biotechnology in countering biological and toxin weapons is addressed in seven principal sections:

1. The wider political and economic context
2. Enabling technologies for BTW agent detection
3. The applicability of biotechnological methods for BTW agent detection on the battlefield, in a terrorist incident and in an inspection environment
4. Pre-exposure medical countermeasures to BTW agents
5. Diagnosis and identification of BTW agents
6. Post-exposure treatment and decontamination of BTW agents
7. The contribution of biotechnology to strengthening international norms against BTW agents.

In this overview, each of these principal sections is considered in turn and followed by a summary based on the principal conclusions reached during the Advanced Research Workshop in the final Workshop session.

1. The Wider Political and Economic Context

The first section addresses the role of biotechnology in countering BTW agents in the wider context of the political and economic dimensions of countering these agents and of strengthening the BTW prohibition regime. Here, Harald Müller sets the scene by considering three scenarios and two dilemmas which scholars and policy makers alike have to confront. These scenarios relate to (1) protecting troops in the field, (2) preventing a disaster for the civilian population in the wake of a terrorist attack, and (3) supporting the verification system of a strengthened Biological and Toxin Weapons Convention (BTWC[*]). In each scenario, he considers the issues of scope, i.e. the range of possibilities that a detection/identification/countermeasure system has to cope with, and time, i.e. the speed with which such a system will be required to perform. He concludes by addressing the dilemmas between preventive efforts by industry and the armed services on one hand, and the demands of cooperative, multilateral arms control measures on the other.

Graham Pearson starts his overview of the prospects and requirements for countering biological warfare with an assessment of the nature of the biological weapons (BW) threat. Following this he traces first international developments – such as revelations about the Iraqi and the past Soviet offensive BW programs and the attempts by the Japanese Aum Shinrikyo cult to use BW as well as chemical weapons (CW) – which have heightened attention and risk perceptions in relation to the acquisition and use of BTW agents by states and sub-state actors alike. To alleviate the increasing political concerns that have been voiced repeatedly, he sets out a four-tier web of deterrence made up of 1) a comprehensive prohibition of BTW acquisition and use through international treaties and national legislation; 2) broad controls ensuring that dual-use materials or equipment are used only for permitted purposes; 3) broad band protective measures, both active and passive, and 4) determined national and international responses to non-compliance with the BW prohibition. He concludes by stressing the importance of the early conclusion of the BTWC compliance protocol as a means to strengthen the BTWC.

The following chapter by Alexander Kelle and Jiri Matousek addresses the lessons that can be learned from establishing the chemical weapons control regime for the ongoing negotiations of the Protocol to strengthen the BTWC. To this end they establish in a first step the commonalities and differences between chemical agents on one hand and BTW agents on the other, discussing *inter alia* the taboo against CBW, the dual-use character of most materials and technologies involved, and the prospects for protection against CBW agents. They then examine the three central compliance measures contained in the Chemical Weapons Convention – declarations, routine and challenge inspections – and discuss the applicability of parallel measures in the BTW context and analyze the state of BTWC protocol negotiations with respect to these three central compliance measures. They conclude by pointing to the 25 year history of the BTWC and caution that the future protocol must not contain any limitation of the BTWC itself, especially its Article I, and that the incorporation of some flexibility in the protocol so as to adapt it to future technological developments will be essential for the protocol's success.

---

[*] The terms "BTWC" and "BWC" are used interchangeably throughout the book.

The remainder of the first section addresses industry involvement and concerns about the future BTWC protocol. René van Sloten presents the views of the FEBC, the Forum for European Bioindustry Coordination, on a protocol to the BTWC which appear to be shared by industry in other parts of the world. As he explains, FEBC is very interested in being involved in the Geneva negotiation process. This interest stems from the recognition that many of the R&D and manufacturing technologies and materials used in the biotech industry could potentially be misused by others to produce biological weapons. FEBC considers it essential that measures strengthening the BWTC ensure that any misuse of microbiology and newly emerging biotechnologies for weapons of mass destruction (or for terrorist purposes) is prevented, without impairing legitimate use and continuing development of biotechnology in academia and industry. In this context, FEBC's concerns relate to the scope of the compliance measures (declarations and inspections) as well as to the protection of confidential business information (commercial and proprietary information). Industry's concerns also include the costs of likely disruption to their activities resulting from an intrusive inspection, and the potential bureaucracy of dealing with complex declaration forms. In conclusion, van Sloten emphasizes that a proper dialogue between the States Parties and industry is needed to ensure that the right balance is struck between the interests of both sides.

In the subsequent chapter Lev Sandakhchiev and Sergey Netesov discuss ways and means of strengthening the BTWC through R&D restructuring. In their outline of the general characteristics of the case they present – the State Research Center of Virology and Biotechnology "Vector" – they stress that the center has established an advanced scientific capability in which research on highly dangerous infectious material can be carried out at Biological Level-3 (BL-3) or BL-4 containment and theoretical and experimental studies in the areas of genetic engineering, biotechnology and aerobiology can also be conducted. Following dramatic decreases in defence-related projects at the Center an extensive R&D restructuring process was begun. In addition, production facilities were set up for the production of, *inter alia*, vaccines and diagnostic kits. Product sales already account for almost three quarters of "Vector's" total income. Still, these efforts remain insufficient and continue to leave institute staff badly underpaid. As a remedy the creation is proposed of International Centers for the study of viral and bacterial pathogens. One critical feature of such Centers should be the inclusion of all high containment capabilities and supporting facilities in order to alleviate concerns about illicit activities in such facilities.

In the final chapter of the first section John Compton examines the impact of the dissolution of the Soviet Union and the subsequent introduction of a market economy on the Russian Biotech industry. He points out that in the early 1990s Russian science in general had suffered enormous setbacks in terms of government funding and support which in turn had led to a severe loss in personnel, and even the closing of some institutes. Many hopes for the commercialization of Russian biotechnology capabilities did not materialize and several factors underlying the reluctance of foreign investors are discussed. Addressing some of these factors – such as the training of institute staff to meet international Good Laboratory Practice (GLP) and Good Manufacturing Practice (GMP) – will also aid in meeting the additional challenges emanating from a future BTWC compliance protocol.

2. Enabling Technologies for BTW Agent Detection

Section II focuses on the technologies which enable biotechnology to be a useful and effective tool-kit in BTW agent detection. In his discussion of antibody-based methods Peter Sveshnikov notes that monoclonal antibodies have been produced against a variety of viruses, bacteria and toxins. These antibodies have been used in different types of immunoassays for the detection of pathogens, with emphasis on increasing sensitivity and specificity as well as reducing assay time. The immunofiltration method that has been developed allows the detection of antigens and antibodies in the nanogram concentration range within 15 minutes. The method allows the analysis of large volumes of diluted samples and continuous monitoring of pathogens.

The contributions by Dario Leslie and by Alexej Prilipov and Kathryn Nixdorff focus on DNA methods. Leslie points out that although a number of categories of microbial components have been used for detection targets via biochemical or immunological assays, the advances in genetics technology over the last few decades offer perhaps the best potential target of all - genetic material. This material contains information that identifies a parent microbe unambiguously and is also chemically stable. Genetic assays are used widely in the medical realm. Research is underway to enhance sensitivity by several orders of magnitude and greatly increase the speed of response. This will help overcome current difficulties in using genetic assays to counter BTW agents and will offer the prospect of a range of practical BW agent detectors for use on the battlefield, as well as in support of the BTWC.

Prilipov and Nixdorff point out that the polymerase chain reaction (PCR) is a powerful biotechnology tool for the identification of microorganisms, which it can identify at the level of groups, genera, species, and even strains. False negative results may be obtained from environmental samples due to inhibitory substances, and various methods are used to concentrate and purify samples. Multiplex PCR systems have been very successful in distinguishing individual variation at highly polymorphic genetic loci and identifying species- or strain-specific variations in microorganisms. A profile based on a set of conserved genes is useful for identifying and tracking the global spread of pathogens. Rapidly evolving genes on the other hand are useful for fine-scale epidemiological investigations detecting genomic differences among microbial isolates during an epidemic. They conclude by expressing the expectation of continued developments in PCR equipment towards compactness and automation in the near future, thereby increasing the prospects for application of such equipment in a number of BTW agent detection scenarios.

J.J. Valdes and James Chambers point out that biosensors – devices that combine biological materials with electronic or optical microsensors which are capable of responding to biological or chemical parameters – so far have a limited range of applications because their selectivity and sensitivity depend on molecular recognition. In their discussion of recent developments in this area Valdes and Chambers touch upon evanescent wave sensors, light addressable potentiometric sensors, cytosensors, and solid state optical sensors. They conclude with a prognosis that future biosensor systems will share three characteristics: they will be reagentless and solid state, microminiaturized, and capable of learning patterns of recognition, which will allow them to modify future responses.

3.   The Applicability of Biotechnological Methods in Different Environments

The third section covers the applicability of biotechnological methods for BTW agent detection in three different environments, i.e. on the battlefield, related to terrorist attacks, and in an inspection environment. Henri Garrigue outlines the characteristics of the threat posed by agents on the battlefield and reviews the strategy of detection appropriate for such agents. A brief discussion of different types of detection requirements then leads Garrigue into a detailed account of standard techniques presently available for detection, how these are presently operationalized, and what might be developed in the future.

Gary Eifried discusses the role of biotechnology in detecting bioterrorist attacks. He argues that capabilities required of biological agent detectors in terrorist incidents can be significantly different from the capabilities required for use on the battlefield. Four specific functions of detection systems have to be distinguished: warn, confirm/deny, identify, and clear. An outline of requirements is presented, and current and developmental technologies are evaluated against these requirements. The results show that it is unlikely that a single technology will suffice, and that public safety agencies will have to rely on a suite of devices and technologies to meet biological detection needs.

Graham Pearson considers the various inspection scenarios being addressed in the negotiations on the BTWC protocol. He examines the role that sampling and analysis – with detection and identification of BTW agents – can contribute to an effective regime in the light of experience gained in other regimes ranging from the UN Secretary-General's investigations of alleged use to the UN Special Commission (UNSCOM) on Iraq and under the Chemical Weapons Convention (CWC). He concludes that sampling and analysis using validated analytical techniques in internationally accredited laboratories is an essential tool in an inspection environment, even though it may be used only infrequently.

4. Pre-exposure Medical Countermeasures to BTW Agents

The fourth section deals with the pre-exposure medical countermeasures to BTW agents. Kathryn Nixdorff addresses the possibilities and limitations of vaccinations and points out that depending on the types of vaccines and their administration, very different types of immune responses will be achieved. The ways in which antigens are delivered, taken up and processed by antigen-presenting cells need to be taken into account when designing a vaccination strategy. In her assessment, the use of DNA vaccines is a promising development. Although vaccines do not afford a perfect defense against BTW agents, they can provide effective protection against many infectious microorganisms. Consequently, the possession of a good vaccine against a particular agent can act as a deterrent to discourage an aggressor intending to use that agent.

Netty Zegers *et. al.* then present their work on the development of an oral vaccine against one BTW agent that is widely considered to be a likely candidate in future offensive BTW programs – *Bacillus anthracis*, which causes the disease anthrax. Starting from the recognition that current vaccines for anthrax are far from ideal, as they are probably not fully effective against all strains of the bacterium, they advocate the development of a new generation of anthrax vaccine. The use of *Lactobacillus* as a vector for expression of heterologous proteins from pathogens can serve as a safe system, which can be given orally to target the vaccine to mucosal surfaces, where most infections begin. The first steps in the development of an oral carrier vaccine are outlined in which the gene encoding the Protective Antigen (PA) of *B. anthracis* was cloned into a homologous vector and subsequently transformed to *Lactobacillus casei*. Expression of the PA in *L. casei* was achieved and PA-specific antibody responses could be obtained upon immunization with a cell lysate of *L. casei* expressing PA.

Jack Melling, in his analysis of vaccine production for protection against BTW agents, starts from the recognition that advances in biotechnology have enabled the production of vaccines against an increasing number of pathogens. Yet, the potential number of BTW threat agents presents a major challenge to the use of vaccines for protection. The response time to produce quantities of a vaccine which has already been fully developed and licensed can be measured in months. For development and approval of a new vaccine the timescale is at least an order of magnitude greater. This gives a significant advantage to a potential aggressor. Thus, for a vaccine strategy to be effective, both stockpiling and immunization of personnel at risk well in advance of any attack would be essential. In view of the large and still growing number of potential BW agents, Melling strongly advocates research into both improved physical protection of at risk personnel and additional means of mobilizing immune responses by a major effort to understand and apply non-specific immune stimulation.

5.   Diagnosis and Identification of BTW Agents

In the fifth section the role of biotechnology in diagnosis and identification of BTW agents is discussed. In this context György Berensci and Gábor Faludi give an overview of the identification of symptoms from a medical point of view. They first discuss BTW-relevant products and uses of biotechnology – such as genetically manipulated organisms for use at different stages in industrial production, and technology to apply "in process control" to biotechnology procedures – that may pose a risk through their potential offensive military application. They then analyze changes in diagnostics and laboratory technology for the detection of small, specific molecular differences before assessing sources of particular concern in the area of virology.

Following this David Franz gives an overview of medical countermeasures to BTW agents and how these countermeasures can benefit from advances in biotechnology. With respect to a central element in any biodefense program – vaccines – Franz argues that advances in biotechnology are facilitating the development of vaccine candidates in the laboratory and eliminating constraints such as the requirement for biocontained production laboratories. These new technical options and methods could also reduce regulatory compliance difficulties in the future. Concerning antibiotics to counter bacterial agents, Franz points out that molecular biology can be an important tool in helping us to understand and overcome the mechanisms of microbial resistance to antibiotics. Viral genomics, on the other hand, will increase our understanding of the target and thereby focus the development of new antiviral drugs. He then points out additional benefits to be derived from biotechnology in the areas of antibody preparations and rapid diagnostics, before concluding that breakthroughs in biotechnology will on balance strengthen defense against BTW rather than its misuse for offensive military applications.

6.   Post-exposure Treatment and Decontamination of BTW Agents

The sixth section contains three contributions on the role of biotechnology in the post-exposure treatment and decontamination of BTW agents. First, Sven-Åke Persson addresses the treatment of mass casualties following a BTW attack under worst case assumptions. He reminds us that worst case scenarios differ substantially depending on where an attack with BTW agents is assumed to have taken place. An attack is likely to have much worse consequences in a less developed country with poor infrastructure and limited health care systems than in a developed society with a good infrastructure. Following this he sets the scene for and discusses three worst case scenarios, two of them in a developed country and one in a less developed country, pointing out the differences in possible responses.

Sergey Netesov then presents possible post-exposure prophylactic measures against viral BTW agents. Viruses, as disease causing agents, have some features which distinguish them from other microorganisms not only by shape or size but also by their mechanism of reproduction. They have to use intracellular structures and enzymes for their replication and therefore the critical stage is the stage at which they penetrate the cell. The main, if not the only, way to defeat a virus is to attack it before it enters the living cell, whereas bacteria may be attacked throughout their life cycle in the infected macroorganism. These peculiarities of viruses limit the possible ways of preventing disease after infection, but it is still possible if drugs or vaccines are used during the significant time interval between the moment of infection and the pathogenic stage. Netesov analyses the use of chemical drugs, immunoglobulins and vaccines as post exposure prophylactic measure against this background. The examples he uses include hepatitis A virus, rabies virus, tick-borne encephalitis virus, influenza virus, and Ebola virus.

Finally, R. Dierstein, H.-U. Glaeser and A. Richardt report on efforts to develop biotechnology-based decontaminants for CBW agents. In the light of existing decontamination methods for CBW agents which rely on thermal and/or chemical principles and are mostly difficult to handle, the potential of biotechnological methods to provide non-toxic detoxifying and disinfecting components is presently being investigated. Several detoxifying enzymes for nerve and blister agents, toxin and infectious agents have already been identified. Further research supports the conclusion that biotechnological approaches may become fitting alternatives to existing protective measures against CBW and may lower the logistic and operational burden involved in the application of decontaminants. Such approaches are less hazardous, less corrosive, potentially applicable to the skin, and environmentally compatible.

7. The Contribution of Biotechnology to Strengthening International Norms Against BTW Agents

The final section of the book assesses the impact of biotechnology on politico-military and normative efforts to counter BTW agents. Malcolm Dando, examining the politico-military responses, first looks briefly at the nature of the threat and the range of means available to counter it. This leads him on to a consideration of the biotechnology revolution which – as he reminds us – is only the latest of a series of scientific/technological developments which have had an impact on the evolution of biological weapons programs over the last century. His analysis of politico-military responses is then set within that historical framework. Dando draws here on the web of deterrence developed by Graham Pearson. Accordingly, arms control, sanctions, and the use of armed force are each considered in turn for their potential. He concludes by emphasizing the need to achieve an increasing regulation of the military aspects of biotechnology.

Finally Alexander Kelle focuses on the relationship between biotechnology and the development of norms against BTW agents. He identifies three main areas where modern biotechnology is of particular relevance to the development of norms and rules against BTW agents. He argues that the way in which the biotechnological and pharmaceutical industries insist on the protection of their commercial proprietary information, together with the apparent readiness of some governments to agree to the demands of their respective industries, threatens to lead to a Protocol which does not strengthen or develop further norms against BTW agents but might instead lead to the weakening and undermining of the existing norms. Kelle then addresses the dual-use problem and the offense–defense dichotomy in relation to the establishment and setting up of normative structures which can guide behavior in efforts to counter BTW agents. In this context he places special emphasis on what individuals or groups involved in these two areas can do to act in accordance with, complement, and strengthen the norms of the regime.

8. Summary Conclusions

The NATO Advanced Research Workshop that formed the basis of this volume has demonstrated that modern biotechnology not only makes it easier to produce the quantities of biological and toxin weapon agents necessary to carry out an attack but also increases the prospects of countering BTW agents. In addition, whilst genetically modified agents can be produced, it seems probable that the unmodified traditional BTW agents will continue to present the greatest danger as it is these agents that have been proved by all means short of actual use in past offensive biological weapons programs. Genetically modified agents will not have been subjected to such extensive trials and there will thus be less confidence that they will be effective if used in weapons. It is therefore recommended that measures to counter biological and toxin weapons be focussed primarily on the traditional unmodified agents although the danger from genetically modified agents cannot be ignored.

The key elements of protective measures against biological and toxin weapons attacks are:

8. Hazard Assessment: An evaluation of the likely agents and the concentrations that may be encountered at various distances downwind from the point of attack.
9. Detection: The rapid and accurate detection of a BTW agent concentration in the atmosphere to enable warning and alerting of those at risk.
10. Physical Protection: The adoption of measures to prevent inhalation of a harmful concentration of the biological or toxin weapon agent.
11. Identification and Diagnosis: Rapid identification and diagnosis so that appropriate medical countermeasures can be taken.
12. Medical Countermeasures: These may be taken prior to an attack (prophylaxis) or after exposure (treatment). There is a particular requirement for products licensed for both safety and efficacy.

Biotechnology has been vital for the development of detection techniques, for identification and diagnosis, and for medical countermeasures. Indeed, it would be accurate to say that detection, identification and diagnosis, and medical countermeasures would be impossible without biotechnology.

It consequently follows that the recent advances in biotechnology offer a real opportunity for effective counters to biological and toxin weapon agents to be developed. As biological and toxin weapon agents occur in natural outbreaks of disease or intoxinations, there are **both** civil and military requirements for counters to biological and toxin weapon agents.

It is useful to consider what the requirements are for warning/alerting, identification, diagnosis and medical countermeasures in a number of different scenarios:
- Battlefield/Attacks on military targets
- Terrorism incidents
- Inspection environment
- Civil outbreaks of disease

The requirements are summarized in the following table:

Scenario	Warning/ Alerting	Identification	Diagnosis	Medical Countermeasures
Attacks on military targets	**Particularly important**	**Rapid** Identification	Required	Required
Bioterrorism	--	Required	**Particularly important**	Required
Inspection environment	--	Required	--	--
Civil outbreaks	--	Required	**Particularly important**	Required

In considering attacks on military targets, warning/alerting is of particular importance because the detection of the attack followed by warning of military resources that may be subject to the attack is a central element of the protective stance. Identification needs to be rapid as the danger posed by the attack on a military target depends on the particular agent used. Diagnosis is required in order to devise the most effective medical countermeasures to the attack on military personnel.

For bioterrorism, much depends on the particular terrorist incident. If a BTW weapon has been used without warning then diagnosis is particularly important as the first manifestation is likely to be the symptoms of those exposed. If the weapon has not been employed, then identification of the nature of the agent is required. Medical countermeasures are also required.

In the inspection environment, there should be no situation in which the inspection team is subjected to an unexpected attack. The safety of the inspection team should be guaranteed and the principal requirement will then be for identification of the nature of any material found.

Finally, for civil outbreaks of disease, the first manifestation is again likely to be the symptoms of those exposed. The agent needs to be identified and medical countermeasures will be required.

Overall, it is evident that biotechnology is vital for the countering of biological and toxin weapons and that, without biotechnology, detection, identification and diagnosis, and medical countermeasures would be next to impossible. It is also clear that the recent advances in biotechnology offer a real opportunity for the development of effective counters to biological and toxin weapon agents. As biological and toxin weapon agents occur in natural outbreaks of disease or intoxinations, there are **both** civil and military benefits from the use of biotechnology in providing effective counters to biological and toxin weapon agents.

DEALING WITH A HEADACHE: THREE SCENARIOS AND TWO DILEMMAS

HARALD MÜLLER
Peace Research Institute Frankfurt
Leimenrode 29
D-60322 Frankfurt/Main
Germany

1. The Virtual Threat

Biological warfare in its rich variations, though largely a non-event so far, has rightly become a central concern of the security and arms control communities. In the history of warfare, this is a strange and unusual phenomenon. In the normal course of events, means of warfare have attracted the highest degree of attention only after serious experiences with their employment in battle have revealed the magnitude of the problem. The evolution of thinking on the other two type of weapons that are dubbed "weapons of mass destruction" illustrates this point.

While the Hague conferences made a weak attempt to curb chemical warfare, only the horrors of the killing in the trenches of World War I convinced the leading states that banning the use of chemical weapons was the right thing to do. It took another seventy years, and the specter of widespread chemical warfare raised by the Iraqi employment of poison gas in the war against Iran, to force the international community to accept a complete ban on these weapons [1].

Nuclear weapons proved their devastating power through the mass fatalities in Hiroshima and Nagasaki and thereby gained pride of place as – in the words of the late Bernard Brodie – "the absolute weapon" in postwar strategy and arms control. In addition, observers have noted that a virtual taboo emerged as an apparently effective norm on the use of these weapons even in highly asymmetric conflicts where deterrence played no evident role.

The same is not true for biological weapons. Apart from the experiments by Japanese occupation forces in China and some minor terrorist incidents, there is no clear-cut, evident employment of this weapon in practice, and certainly not on a large scale. Yet, it is now at the center of military planning and arms control negotiations.

1

A. Kelle et al. (eds.), The Role of Biotechnology in Countering BTW Agents, 1–8.
© *2001 Kluwer Academic Publishers. Printed in the Netherlands.*

The reason for this is to be found in three characteristics of biological weapons: they are relatively easy to develop and produce, their production and stockpiling is difficult to detect from the outside, and the killing potential – given a successful dispersion mode and appropriate environmental conditions provided – is absolutely horrifying. Security planners thus have no choice but to consider different scenarios and try to develop countermeasures. Arms controllers are challenged to consider ways of developing methods to reduce the danger of actual development under non-detection (thereby providing both a deterrent to the perpetrator and reaction time for the international community), though we must accept that such a system can never be perfect.

In this chapter I will try to set out the task before us in more detail by looking at the three quite different tasks that biotechnology is required to contribute to: protecting troops in the field, preventing a disaster for the civilian population in the wake of a terrorist attack, and bolstering the verification system of a strengthened Biological and Toxin Weapons Convention (BTWC). In all three cases, I look at the issues of scope and time, understanding scope as meaning the range of possibilities that a detection/identification/countermeasure system has to cope with, and time as the speed with which the system will be required to perform. At the end, I will point to the inevitable dilemma between preventive efforts by industry and the armed services and the demands of arms control under the auspices of the BTWC.

2. Facing a BW-armed Enemy

Let us begin by looking at the requirements for protecting troops in the field, most likely during an out-of-area operation under a UN mandate. If there is a suspicion that the target of the mission possesses biological weapons, there must be the possibility of the timely detection of employed agents, their speedy analysis and identification, and surge capability for well-targeted vaccination, quite apart from defensive protective gear that must be deployed with the forces. If these capabilities cannot be provided, there will be an increasing risk that international order will be weakened and that more and more countries will feel forced – or encouraged – to give weapons of mass destruction a prominent role in their own national military postures.

A crucial issue is the scope of detection/analysis capabilities. If there are good reasons to assume that the enemy is highly likely to dispose only of well-known, "conventional" agents, detection/analysis capabilities can be tailored to a relatively narrow range of agents, with a high probability of correct identification; in that case, the problems should not be insurmountable. It should be possible either to deploy sufficient vaccines for the deployed units as a matter of routine logistics, or to have mobile laboratories with the production potential to produce complementary vaccines in a short period of time. We are talking about a number of patients in the low hundred thousands, presumably, and about a few hours or at most days during which the work has to be accomplished.

The situation gets considerably more difficult if we are facing an opponent with a sophisticated bioindustrial infrastructure and considerable experience in biological warfare development. In that case, the possibility that troops will encounter unknown design BW agents cannot be excluded, particularly as time goes by and knowledge of the possibilities of genetic combination becomes ever more sophisticated. The requirements for analysis and identification are daunting, and certainly not to be solved with the help of very limited land-mobile laboratories. It would be necessary to follow many analytical tracks in parallel in the hope of being able to determine the nature of the agent quickly. The solution might be to have a fast logistic ship equipped as a defensive, large-scale BW laboratory, including vaccine production capabilities, that would be deployed with the troops from the outset. The alternative is to use home laboratories in an emergency mode, flying the agent there immediately after detection. But this is very much a matter of timing, as the site of the military mission might be fairly remote from the nearest friendly laboratory, and the flying time may eat up the precious time needed to counteract the threat. In addition, there might be some political resistance against introducing an unknown BW agent into home countries. For these reasons, it would certainly be preferable to have sufficient capability in the theater of operations.

3.   The Terrorist Threat

The second scenario concerns a terrorist attack with biological weapons [3,4,5]. To prevent this in all places and at all times is, of course, impossible. It is equally impossible, most probably, to have adequate protection prepared everywhere for all contingencies. One has to think of countering the most probable and most threatening cases. Terrorist groups are very likely to work with traditional agents and to focus only on a few sites – most likely a single one. Since they will select such sites for the largest number of fatalities (after having opted for weapons of mass destruction in the first place), defensive precautions should be concentrated in the ten largest cities of a country, with some planning for how to draw on these assets should an attack occur in an unexpected place.

These cities should have emergency plans for this class of events. Doctors and health workers should be prepared for the mass action – treatment of victims and preventive measures for those not yet affected – at short notice. Vaccines should be stored for the most common agents. Those required to deal directly with the problem – police, employees of the health service – should have long-term preventive vaccination, if feasible. The police should dispose of special units trained in agent detection and sampling. There should be a designated emergency laboratory for analysis and identification that is able to start work at a moment's notice. It is quite clear that this proposal means an upsurge in civil defense efforts, which will surprise many after the end of the East-West conflict. It needs some investment, but, at first glance, not terribly much since it could draw mostly on civilian assets that are there anyway. This scenario involves the typical dilemma of a low-probability, but grave-consequences contingency; political authorities have to take precautionary measures if they wish to live up to their responsibilities. But the public may not be convinced that these measures make sense until the first fatal incident happens, at which point it would be too late.

The hypothesis that the terrorist type of threat will be confined to traditional agents hinges on the assumption that the terrorists have no systematic link to a state that possesses the most modern, sophisticated BW facilities. If such a link exists, the possibility of encountering new and unknown agents would be greatly enhanced, and the challenge for civil defense would be correspondingly greater. The requirements for laboratory work on analysis and identification would multiply, as would the need for the quick production of vaccines and medicine for treatment. As we are talking about several hundred thousand up to many million people, the logistics of production and distribution would become a real headache. It might be wise to consider planning for such a contingency only after the more straightforward measures for the traditional-agent scenario are in place and well exercised.

Preparing technically for this contingency is also important for political reasons. The degree of reassurance provided through these precautions may help to reduce the strong impulse to compromise on citizen rights and liberty in order to prevent this huge threat from being realized. The threat of biological terrorism alone, mounted to destroy our societies and their democratic systems, could otherwise achieve its sinister aim eventually by default.

4. The Arms Control Mission

For the arms control objective the time requirements are much more relaxed. The delay between sample-taking and the completion of the analysis would be determined by the legal content of the forthcoming protocol. Of course, in the case of a sample taken in the context of a clarification visit or even a challenge inspection the delay should not be too great as the political situation would be tense until the issue was resolved. But no human lives would be at stake, which would mean the urgency was less great than in the contingencies discussed above. On the other hand, reaching valid conclusions and conveying them publicly to the international community is a much more delicate and difficult task than for the other two scenarios, which are largely national-political, or multinational-military, matters. Institutional and procedural provisions must be very thoroughly calibrated to do the job in a convincing and thus effective way.

It should also be noted that the requirement to produce medicine and vaccine quickly would no longer be a necessity. On the other hand, the scope of the analysis would be virtually unlimited, since the possibility of unknown and design agents would loom large. While there are suggestions that analysis should be confined to a range of known agents that should be listed explicitly in the protocol, it is hard to see how this proposal can be squared with the comprehensive language in Article I of the BTWC. In addition, the confidence-building and, in the case of the detection of a real breach, security value of the Convention and its forthcoming protocol would be largely lost if the possibility of detecting non-listed, new agents were to be foreclosed.

5. The Arms Control versus Commercial Interest Dilemma

This leads us, however, to the two dilemmas we are facing in the arms control/BTW threat context. The first dilemma emerges from the legitimate interest of industry, but also university research, in seeing intellectual property and commercial interests protected against unfair competitive practices that would emerge from an undue "leakage" of knowledge gained through sampling and analysis. Biotechnology is, of course, a field with an enormous innovative potential and it will remain so for a considerable time. New insights and, consequently, new substances of commercial value are developed by the day. If a company admits sampling on its premises, it runs a certain risk that laboratory analysis elsewhere will reveal the character of a new product and may then lead to the premature appearance of competitive products in other countries. It is not surprising, therefore, that many in industry have strong reservations about an unlimited authority for an international organization to sample and analyze, and plead for much more constrained modes of transparency and verification. Such constraints, in turn, could hamper the construction of a verification system with minimum credibility.

Can this dilemma be solved technically? Would it be possible to analyze a sample first for known agents; if the finding is negative, to probe its lethality; if the substance is found to be lethal, to assess the indications that the intended use might be incompatible with Article I of the BTWC, and only if this assessment leads to alarming suspicions, apply broad-scale analysis to the sample, while the sample would be disposed of forthwith if it did not show lethal characteristics?

6. The Arms Control versus Self-Protection Dilemma

The second dilemma concerns the first two scenarios and the boundary between permitted and prohibited activities. It might be possible to draw the line very clearly for the civil defense measures discussed in the second scenario. They must be reported and could be inspected to verify their scope and compatibility with Art. I of the BTWC. As they would normally be under civilian control, it should be possible to distinguish them clearly from clandestine preparation for active biological warfare.

This distinction is almost certainly much harder to draw in the first, the military defensive preparation, scenario. The procurement of large-scale, mobile laboratory facilities with the capacity for unknown-agent analysis and emergency production capacity for countermeasures, the recruitment and training of the appropriate specialized manpower, the exercising of troops for emergency defensive responses to an acute threat – all this overlaps with requirements for offensive biological warfare. Is it possible to draw the distinction on technical grounds? To identify observable signatures, without very intrusive observation, that would allow us to distinguish the permitted from the prohibited? I have doubts, but remain open to persuasion.

The solution, in that case, may lie in institutional and human arrangements that would need to be highly innovative. Permanent neutral observers attached to the BW defense units might be in a position to judge from the daily operating routines whether the purpose of the preparations was offensive or defensive. Such observers must be recruited from countries friendly enough to the inspected parties as to be acceptable as attachés, and still detached enough to be seen by the potential enemies of the inspected country as sufficiently trustworthy. Rules of observation would have to walk the tightrope between yielding enough insight to build confidence yet not compromising the inspected party's national security. This would not be easy.

7. Conclusion

What has become clear is that the challenge for the application of technology in the fields of detection, analysis/identification, and the development of countermeasures is daunting with regard to all three scenarios. The requirements in terms of time, scope, and – in the case of the civil population – numbers, but most particularly the combination of these requirements in some contingencies, are very tough and distinguish this field sharply from the requirements usually faced by civil research and industrial development. That this has to be solved in the face of two serious political dilemmas does not make the task any easier.

8. References

1. Price, R. (1997) *The Chemical Weapons Taboo*, Cornell University Press, Ithaca, New York.
2. Tannenwald, N. (1999) The Nuclear Taboo: The United States and the Normative Basis of Nuclear Non-Use, *International Organization* **53(3)**, 433-468.
3. Carter, A., Deutch, J., and Zelikow, P. (1998) Catastrophic Terrorism, *Foreign Affairs*, **77(6)**, 80-94.
4. Roberts, B (ed.) (1997) *Terrorism with Chemical and Biological Weapons: Calibrating Risks and Responses*, Institute for Defense Analysis, Alexandra, Va.
5. Falkenrath, R.A., Newman, R.D., and Thayer, B.A. (1998) *America's Achilles' heel: nuclear, biological, and chemical terrorism and covert attack*, MIT Press, Cambridge, Mass. (CSIA studies in international security).

COUNTERING BIOLOGICAL WARFARE: AN OVERVIEW

GRAHAM S. PEARSON
Visiting Professor of International Security
Department of Peace Studies
University of Bradford
Bradford, West Yorkshire BD7 1DP, United Kingdom

1. Introduction

The past five years have seen significant developments in the regimes for the prohibition of chemical and biological weapons with the entry into force of the Chemical Weapons Convention on 29 April 1997, the progress being made by the Ad Hoc Group of the States Parties to the Biological Weapons Convention in negotiating a legally binding instrument to strengthen the Convention, and the lifting of reservations to the Geneva Protocol. However, the dangers to both international and national security from both chemical and biological agents and materials continue. Indeed, biological weapons currently pose the greatest threat of all weapons of mass destruction as they are the easiest to acquire, have the weakest regime and yet have effects comparable to nuclear weapons. The counter to such weapons of a web of deterrence comprising arms control, export controls, protective measures and determined national and international responses will be of vital importance in ensuring future security, peace and prosperity.

This chapter addresses the international developments during the last decade leading to the current situation and the increasing political concern about biological weapons. It then considers how to counter biological warfare, with particular attention being given to the role of biotechnology. It concludes by considering what needs to be done in order to build national and international security.

2. The Nature of Biological Warfare

Biological warfare is the use of disease as a weapon of war – whether against humans, animals or plants. It is essentially deliberate disease [1] caused by the

9

A. Kelle et al. (eds.), The Role of Biotechnology in Countering BTW Agents, 9–31.
© 2001 Kluwer Academic Publishers. Printed in the Netherlands.

deliberate dissemination of pathogens to cause disease amongst humans, animals and plants. Disease is caused by the entry of small numbers of living microorganisms into the target population of humans, animals or plants; these then multiply and after an incubation period, the symptoms of the disease become apparent. These symptoms in the case of some microorganisms are the result of the production of toxins, non-living toxic chemicals, by the microorganisms. Consequently, biological warfare is used to describe the deliberate use of both living microorganisms and toxins to cause harm in humans, animals and plants. Toxins are non-living toxic chemicals produced by living microorganisms; they can also be produced by chemical synthesis and consequently the prohibition in the Biological and Toxin Weapons Convention [2] of the misuse of biological agents and toxins contains the wording "microbial or other biological agents, or toxins whatever their origin or method of production."

There are several types of microorganism that can cause disease – bacteria such as anthrax or plague, viruses such as smallpox or Ebola and rickettsia such as Q-fever. All, if misused, would be covered by the term biological warfare, and all, including any genetic modifications, are covered by the prohibition in the Convention. The principal groups or classes of the pathogens that can cause infectious disease and of toxins which can cause harm through intoxination and could thus be potential biological agents are:

- **Bacteria.** These are the causative agents that produce diseases such as anthrax, plague and tularemia. Although many pathogenic bacteria are susceptible to antibiotic drugs, strains can be selected that are antibiotic resistant and occur naturally. They can be readily grown in artificial media using facilities akin to those in the brewery industry.
- **Viruses.** There are large numbers of viruses that produce diseases such as Venezuelan equine encephalitis. These must be grown on living tissue.
- **Rickettsia.** An example is the organism that produces Q-fever. These are intermediate between viruses and bacteria and must be grown in living tissue.
- **Fungi.** An example is coccidioidomycosis. Relatively few species appear to have potential for deliberate use against humans although many more have potential against plants.
- **Toxins.** These are the non-living products of microorganisms such as botulinum toxin or staphylococcal enterotoxin B, of plants such as ricin from castor beans or of living creatures such as saxitoxin from shellfish.

Consideration of whether a particular pathogen or toxin might be selected as a biological agent requires knowledge of a large number of factors, such as the infective dose, the time to effect, and whether the agent produces a transmissible disease, as well as the method of attack of the target population (inhalation, ingestion, or by an insect vector), the means of dispersion of the agent, the stability of the agent and the practicality of an infective dose being achieved in the target population. Similar considerations apply to toxins, but because toxins are non-living, they cannot produce a transmissible disease. Toxins are much more closely related to chemical warfare agents; unlike some biological agents, toxins can only affect those exposed to the agent.

There are certain characteristics which will make a pathogen more attractive as a BW agent: it will need to be producible, able to infect the target population, generally by inhalation although ingestion is an alternative, have sufficient stability to be disseminated and to survive in the atmosphere until it reaches the target. As with any disease, there will be an incubation period during which the microorganism will multiply in the host, following which the disease will become apparent resulting in incapacitation or death depending on the pathogen selected. This time to effect can be regarded as both a disadvantage and an advantage. Although the absence of an immediate effect negates potential applications in a contact battle, the delayed effect can be used to advantage in attacking fixed targets such as airbases, ports and logistic concentrations. Additionally, the delayed effect means that attribution is difficult and, especially where an endemic disease is used for an attack, it is plausible to both hide and deny the BW attack. The feasibility of biological warfare was demonstrated by all means short of actual use in war in work carried out during the years after World War II and before the opening for signature in 1972 of the Biological and Toxin Weapons Convention.

Because biological agents (other than toxins) multiply in the infected target population, the quantities needed to cause infection are very much smaller than the amounts of chemical agent needed to cause harm – a few biological microorganisms may suffice. Consequently, BW has a significantly larger potential area of effect than have chemical weapons and hence the potential impact of BW approaches that of nuclear weapons and can have strategic effects.

The downwind hazard from a biological weapons attack can extend to a few hundred kilometers if the meteorological conditions are optimum. As the quantities required for BW are small compared to chemical weapons, they can be disseminated by cross–wind dissemination with few if any indications of hostile intent. A simple dissemination system such as one mounted on a single aircraft flying across the wind could be used to produce a line source of 200 km or so long, resulting in an attack of an area of some 200 km wide by 200 km downwind; a vehicle driven across the wind could produce a shorter line source of perhaps 50 km in length attacking a proportionately smaller area. The Office of Technology Assessment of the US Congress, in an evaluation [3] of weapons of mass destruction, showed that a single aircraft attack on Washington DC using anthrax could result in 1 to 3 million deaths; in the same scenario, a one megaton hydrogen bomb would only cause some 0.5 to 1.9 million deaths. Such attacks with biological agents present a serious challenge to personnel in military targets such as naval task forces and assembly areas or to strategic targets. Likewise, if used to attack livestock or plants, a significant effect could be produced.

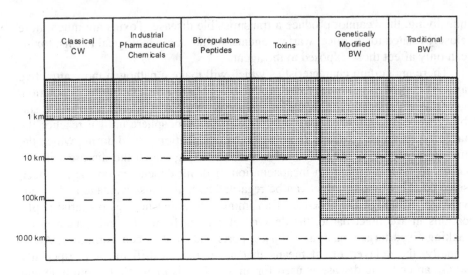

Figure 1. CBW Downwind Hazard

There is now an awareness that biological warfare agents could be produced which have been genetically modified [4]. Whilst this is a potential danger that cannot be dismissed, it needs to be recognized that microbiological organisms are inherently fragile and their ability to survive dissemination into the atmosphere is limited. Consequently, genetic modification to change a particular characteristic may inadvertently have a negative effect on some other characteristic of the organism. It is consequently probable that a would-be acquirer of a genetically modified biological warfare agent would need to evaluate the product through experimental field trials. It is consequently by no means certain that a genetically modified biological warfare agent would pose a greater threat than the original unmodified agent. It is for this reason that the traditional biological warfare agents are seen as presenting the greatest threat – although the possibility that an agent might be genetically modified cannot be disregarded and needs to be borne in mind.

3. International Developments

The past decade has seen increasing international awareness of and concern about biological warfare. The 1990/91 Persian Gulf war saw a real concern among the coalition forces facing Iraq that Iraq had both chemical and biological weapons and might use them against the coalition. Steps were taken to improve the ability of the coalition forces to survive and fight should chemical or biological weapons be used against them. Following the war, the United Nations Security Council set up the United Nations Special Commission (UNSCOM) on Iraq with a mandate to oversee the destruction of Iraq's weapons of mass destruction and to establish an ongoing monitoring and verification system to ensure that Iraq did not reacquire such weapons [5].

At the Third Review Conference in September 1991 of the Biological and Toxin Weapons Convention (BTWC), the States Parties recognized the need to strengthen the Convention and agreed to establish an Ad Hoc Group of Governmental Experts to examine possible verification measures from a scientific and technical viewpoint (this group was known as VEREX) [6]. VEREX met twice in 1992 and twice in 1993, producing a final report which identified, examined and evaluated 21 off-site and on-site potential verification measures [7]. This final report was considered at a Special Conference in September 1994 which agreed to establish an Ad Hoc Group (AHG) "to consider appropriate measures, including possible verification measures, ...to strengthen the Convention..." [8]. This AHG had, by December 1999, met 17 times and had successfully transitioned in July 1997 to the negotiation of a rolling text for the Protocol to strengthen the BTWC. By December 1999, the tenth version of the draft had been produced [9].

In 1992, President Yeltsin of the Russian Federation admitted that the former Soviet Union had, despite being a co-depositary of the 1972 BTWC, continued an offensive biological weapons program. This program was described in 1996 by a UK minister in a statement to the Fourth Review Conference as having been "a massive offensive biological weapons program conducted illegally for years in the Soviet Union" [10]. A trilateral process was set up in 1992 involving the UK, US and Russia to re-establish confidence that the program has ceased. The latest Annual Report of the US Arms Control and Disarmament Agency states, as it has for several years, that the process "has not resolved all U.S. concerns." [11]

The potential danger from terrorist use of chemical or biological weapons was thrown into sharp global focus by the Aum Shinrikyo attack using the nerve agent, sarin, in the Tokyo subway system in March 1995. It subsequently has become clear that the Aum Shinrikyo had also been seeking biological weapons and had attempted to use such weapons although without causing any casualties [12].

The United Nations Special Commission on Iraq (UNSCOM) has completed seven years of hard work in the face of determined Iraqi obstruction aimed at uncovering and destroying Iraq's weapons of mass destruction; in respect of biological weapons, Iraq endeavored to keep its program secret. [13] Thus in June 1995, Iraq had yet to admit to having any offensive biological weapons program. Indeed, it was only on 1 July 1995 that Iraq admitted for the first time its offensive biological weapons program and it was not until after Hussein Kamel defected to Jordan on 7 August 1995 that Iraq admitted its weaponization of biological warfare agents into aerial bombs and Scud missile warheads [14]. It is clear in 1998 that Iraq's "Full, Final and Complete Disclosures" in respect of its biological weapons program is still far from being complete.

4. Increasing Political Concerns

There has been increasing political concern about the dangers from biological weapons. These have addressed both the counters to proliferation of biological warfare to states and the counters to the possible use of biological materials by terrorists. The danger of biological weapons being used against the coalition forces in the Persian Gulf war of 1990/1991 led to the recognition by the Security Council when they met for the first time at Heads of State and Government level in January 1992 to declare that

> "The proliferation of all weapons of mass destruction constitute a threat to international peace and security. The members of the Council commit themselves to working to prevent the spread of technology related to the research for or production of such weapons and to take appropriate action to that end." [15]

Proliferation of weapons of mass destruction was also identified by NATO as a problem requiring special consideration. Following the meeting of the Security Council at Heads of State and Government level on 31 January 1992, two years later, the Heads of State and Government participating in the North Atlantic Council meeting held in Brussels on 10-11 January 1994 in their final communiqué stated that

> "Proliferation of weapons of mass destruction and their delivery means constitutes a threat to international security and is a matter of concern to NATO. We have decided to intensify and expand NATO's political and defence efforts against proliferation, taking into account the work already under way in other international fora and institutions. In this regard, we direct that work begin immediately in appropriate fora of the Alliance to develop an overall policy framework to consider how to reinforce ongoing prevention efforts and how to reduce the proliferation threat and protect against it." [16]

Six months later the Ministerial meeting of the North Atlantic Council on 9 June 1994 issued an Alliance policy framework on proliferation of weapons of mass destruction. [17] This noted the Security Council meeting of 31 January 1992 statement that such proliferation constituted a threat to international peace and security and stated that the January 1994 NATO Summit initiative reflected the fact that there were developments in the evolving security environment that give rise to the possibility of increased WMD proliferation. The Alliance policy framework went on to say that

> "Current international efforts focus on the prevention of WMD and missile proliferation through a range of international treaties and regimes.

The most important norm-setting treaties are the nuclear non-proliferation treaty (NPT), the chemical weapons convention (CWC) and the biological and toxin weapons convention (BTWC). With regard to the NPT, efforts are currently focused on unconditional and indefinite extension of the treaty in 1995, universal adherence to the treaty and enhancing its verification and safeguards regime. For the CWC, the most immediate goal is its rapid entry into force. The BTWC can be strengthened through efforts in the field of transparency and verification. The Allies fully support these efforts."

The Alliance policy framework then stated that "WMD and their delivery means compose a direct military risk to the member states of the Alliance and to their forces" and said that NATO's approach to proliferation will therefore have both a political and a defense dimension. A series of goals were set out for the political dimension followed by a series of goals for the defense dimension.

The political dimension was addressed by a Senior Politico-military Group on Proliferation (SGP) and the defense aspects by the Senior Defense Group on Proliferation (DGP). In December 1994, the communiqué issued by the Ministerial Meeting of the North Atlantic Council said

"Work on the summit initiative of the proliferation of weapons of mass destruction and their delivery means have been taken forward. ...We welcome the progress made in intensifying and expanding NATO's political and defence efforts against proliferation, which remains one of the greatest concerns for the Alliance. ...We welcome the consultation with all cooperation partners in the framework of the NATO and look forward to ad hoc consultations with Russia on proliferation of weapons of mass destruction and their means of delivery." [18]

At the May 1995 Ministerial Meeting of the North Atlantic Council the final communiqué said:

"We attach high importance to the ongoing work inside the Alliance on the summit initiative of the proliferation of weapons of mass destruction and their delivery means. We took note of the report of the Alliance's joint committee on proliferation on the activities in the senior politico-military group on proliferation and the senior defence group on proliferation. We welcome the progress made in intensifying and expanding NATO's political and defence efforts against proliferation which remains one of the greatest concerns for the Alliance." [19]

In June 1997, the meeting of the North Atlantic Council reaffirmed its commitment to address proliferation risks as an integral part of the Alliance's ongoing response to the new security environment. Its final communiqué stated that "While prevention remains our primary aim, we recognize that proliferation of nuclear, chemical and biological (NBC) weapons and their means of delivery ... can pose a direct threat to the Alliance." [20] The Council also specifically welcomed the emphasis being put on "improving protection against biological weapons."

The importance of completing the Protocol to strengthen the BTWC has been stressed by several political developments during 1998. President Clinton in his State of the Union address on 27 January 1998 said "Now, we must act to prevent the use of disease as a weapon of war and terror. The Biological Weapons Convention has been in effect for 23 years. The rules are good, but the enforcement is weak – and we must strengthen it with a new international system to detect and deter cheating." [21] In March 1998 the European Union issued a Common Position that commits the 15 Member States as well as the 14 Associated States to "actively promote decisive progress in the work of the Ad Hoc Group, with a view to concluding the substantive negotiations by the end of 1998, so that the Protocol can be adopted by a Special Conference of States Parties early in 1999." [22]

In addition, Australia announced on 2 March 1998, an initiative to strengthen the BTWC "aimed at fast-tracking the negotiations on a verification system for the Biological Weapons Convention by ... calling for the convening of a high level meeting to inject into the negotiations the necessary political commitment for urgent action, ... to help secure early conclusion to the negotiations." [23]

A Statement by the Non-Aligned Movement and other countries in March 1998 said inter alia that "they will contribute fully to this work [of the Ad Hoc Group] in order to promote consensus on key issues which will facilitate the conclusion of this undertaking in a manner acceptable to all States Parties..." [24]

In May 1998, the Foreign Ministers of the G8 (which includes Japan, Russia and the United States) in the communiqué issued following their meeting in London said that they "are committed to ... the intensification and successful conclusion of the negotiation on measures, includes those for effective deterrence and verification to strengthen the Biological and Toxin Weapons Convention with the aim of the earliest possible adoption of a legally-binding Protocol." [25]

Later in May, the Ministers of Foreign Affairs and Heads of Delegations of the Non-Aligned Movement in their communiqué following their meeting at Cartagena des Indias, Columbia said that

> "The Ministers noted the progress achieved so far negotiating a Protocol to strengthen the Biological and Toxin Weapons Convention and reaffirmed the decision of the Fourth Review Conference urging the conclusion of the negotiations by the Ad Hoc Group as soon as possible before the commencement of the Fifth Review Conference and for it to submit its report...to be considered at a Special Conference." [26]

The Foreign Ministers also expressed their concern about Article X measures saying that

> "Substantive progress in strengthening the application and full operationalization of Article X is **crucial** to the conclusion of a universally acceptable and legally binding instrument designed to strengthen the Conven-

tion. They reaffirm readiness to work with other delegations in order to achieve an appropriate balance in the Protocol." [Emphasis added] [27]

A few days later, President Clinton announced on 22 May 1998 a major US initiative to counter attacks using biological weapons when he said

"We must strengthen the international Biological Weapons Convention with a strong system of inspections to detect and prevent cheating. This is a major priority. It was part of my State of the Union address earlier this year, and we are working with other nations and our industries to make it happen." [28]

The opening day of the June 1998 meeting of the AHG saw an address by the UK Foreign Minister of State speaking on behalf of the EU entitled "Time to Accept the Realities of the Control of Biological Weapons" which stressed that the reality is that such a regime can be achieved. The Minister went on to urge that the impetus towards the early and successful conclusion of the negotiations should be sustained. He noted

"the great importance attached by many Delegations to the provisions foreseen under Article VII of the Protocol [Measures to implement Article X of the BTWC]. I recognise that it will be important to ensure that agreement be reached between the divergent positions on this **crucial** element of the eventual regime. I feel sure that it will be possible to identify measures that will address real needs." [Emphasis added].[29]

Five days later, a Sino-US Presidential Joint Statement on the Protocol to the BTWC was issued in Beijing during President Clinton's visit to China. This gave a useful boost to the work of the AHG by saying

"Both China and the United States support efforts to strengthen the effectiveness of the Convention, including the establishment of a practical and effective compliance mechanism. ...The two sides express their desire to cooperate in the negotiations and work together **to further accelerate an early conclusion** of the negotiations on the Protocol." [Emphasis added]. [30]

The final day of the June AHG meeting saw two political statements. One, by Argentina, Brazil, Chile, Colombia, Mexico and Peru as the signatories of the 1991 Mendoza agreement or of the Cartagena Declaration, stated that "they attach great importance to and remain fully committed to participate constructively in the negotiations of the Ad Hoc Group of States Parties to the BTWC." The other statement by some 29 Western and Eastern States (ranging from Argentina to Canada, the Czech Republic, the EU States, Japan and the United States) stated that

"the international community must pursue the early and successful conclusion of the Ad Hoc Group negotiations as a matter of urgency ... in order to achieve this goal, ... measures to strengthen compliance should include, inter alia, the following elements...."

* Declarations of a range of facilities and activities of potential relevance...

* Provisions for visits to facilities in order to promote accurate and complete declarations...

* Provision for rapid and effective investigations into concerns over non-compliance...

* A cost-effective and independent organisation, including a small permanent staff..." [31]

Since the June/July 1998 AHG meeting there have been further political developments. A Ministerial meeting held in New York on 23 September 1998 for example was attended by 30 Ministers and supported by a further 27 countries. They agreed a declaration that stated:

"The Ministers are determined to see this essential negotiation brought to a successful conclusion as soon as possible....The Ministers call on all States Parties to accelerate the negotiations and to redouble their efforts within the Ad Hoc Group to formulate an efficient, cost-effective and practical regime and seek early resolution of the outstanding issues through renewed flexibility in order to complete the Protocol on the basis of consensus at the earliest possible date." [32]

April 1999 saw the Washington Summit of the NATO Heads of State and Government agreeing in their communiqué that:

"We are determined to achieve progress on a legally binding protocol including effective verification measures to enhance compliance and promote transparency that strengthens the implementation of the Biological and Toxin Weapons Convention." [33]

A month later, on 17 May 1999, the European Union adopted a Common Position [34] that stated that

"Agreement shall be promoted, in particular in the negotiations, on the following measures which are both central to, and essential for, an effective Protocol to strengthen compliance with the BTWC:

– declarations of a range of facilities and activities relevant to the Convention, *inter alia*, so as to enhance transparency,

– effective follow-up to these declarations in the form of visits, on the basis of appropriate mechanisms of random selection, so as to enhance transparency of declared facilities and activities, promote accuracy of declarations, and ensure fulfilment of declaration obligations in order to ensure further compliance with the protocol,

– appropriate clarification procedures supplemented, if need be, by on-site activities whenever there is an anomaly, ambiguity or omission in a declaration submitted by a State Party, which requires such procedures. Appropriate clarification procedures shall also be followed whenever a facility meeting the criteria for declaration ought to have been declared but was not,

- provision for rapid and effective investigations into concerns over non-compliance, including both facility and field investigations,
- establishment of a cost-effective and independent organization, including a small permanent staff, capable of implementing the Protocol effectively,
- provision for specific measures in the context of Article 7 of the Protocol in order to further international cooperation and exchanges in the field of biotechnology. Such measures shall include assistance to promote the Protocol's implementation."

Finally, in December 1999, the North Atlantic Council agreed that "We ... welcome the progress made in the negotiations in Geneva on a legally binding Protocol to strengthen the Biological Weapons Convention by ensuring effective verification measures to enhance compliance and promote transparency. We urge that additional efforts be made to complete the remaining work as soon as possible before the Fifth Review Conference of the BWC in 2001." [35]

There can therefore be no doubt at all about the political momentum and expectation to complete the Protocol to strengthen the BTWC. It is now up to the Ad Hoc Group to address the remaining issues with flexibility.

5. The Web of Deterrence

The counter to the danger of biological weapons, whether possessed by States or by sub-state actors, is through a web of deterrence made up of a series of strands:

- Comprehensive prohibition through international treaties and national legislation establishing the clear norm that development, production, storage, acquisition or use of biological weapons are totally prohibited;
- Broad monitoring and controls ensuring that materials or equipment are used only for permitted purposes thus increasing the difficulty of acquiring materials or equipment for prohibited purposes;
- Broad band protective measures, both active and passive, thereby reducing the effectiveness of biological weapons;
- Determined national and international responses to non-compliance with the prohibition ranging from diplomatic actions, sanctions through to armed intervention, making it clear that acquisition of biological weapons will not be tolerated;

These strands are mutually reinforcing and lead a would-be proliferator to judge that acquisition of biological weapons is not worthwhile. [36,37] A single strand alone will not suffice yet together they make the benefits of CBW acquisition minimal. It is, however, vital to ensure that all the strands are strong. Each is considered in turn.

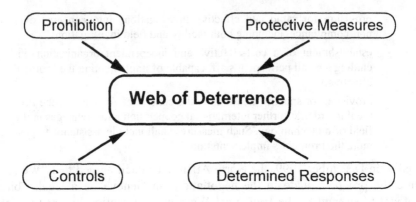

Figure 2. The Web of Deterrence

5.1 PROHIBITION

The norm that the development, production, storage, acquisition or use of biological weapons is totally prohibited is established by the Geneva Protocol of 1925 and the Biological and Toxin Weapons Convention (BTWC) of 1972. The Geneva Protocol prohibits the use of chemical and biological weapons. Although a number of States parties had entered reservations which would have enabled them to retaliate in kind should such weapons have been used against them, these reservations have increasingly been given up in the last few years. Now that the CWC has entered into force, all States still retaining such reservations to the Geneva Protocol should be urged to relinquish them.

The BTWC entered into force in 1975: as of 20 July 1999 it has 143 States Parties and 18 Signatory States. [38] The scope of its prohibition is very broad as Article I contains a general purpose criterion shown in bold below:

> "Each State Party to this Convention undertakes never in any circumstances to develop, produce, stockpile or otherwise acquire or retain:
>
> (1) Microbial or other biological agents, or toxins, whatever their origin or method of production, **of types and in quantities that have no justification for prophylactic, protective or other peaceful purposes;**
>
> (2) Weapons, equipment or means of delivery designed to use such agents or toxins for hostile purposes or in armed conflict." [Emphasis added]

Successive Review Conferences have reaffirmed that the prohibition in Article I applies to all developments by language such as that agreed at the Fourth Review Conference in 1996 that:

> "The Conference, conscious of apprehensions arising from relevant scientific and technological developments, inter alia, in the fields of micro-

biology, biotechnology, molecular biology, genetic engineering and any applications resulting from genome studies, and the possibilities of their use for purposes inconsistent with the objectives and the provisions of the Convention, reaffirms that the undertaking given by the States Parties in Article I applies to all such developments." [Emphasis added] [39]

It is thus clear that all developments in biotechnology are indeed embraced by the prohibition in Article I of the BTWC.

The prohibition of use is not specifically addressed in the body of the BTWC although the Preamble, in stating that "Determined, for the sake of all mankind, to exclude completely the possibility of bacteriological (biological) agents and toxins being used as weapons", makes it clear that the intent is to prohibit use. This intent to prohibit use was reinforced at the Fourth Review Conference by language stating that:

"The Conference reaffirms that **the use by the States Parties, in any way and under any circumstances**, of microbial or other biological agents or toxins, that is not consistent with prophylactic, protective or other peaceful purposes, is effectively a violation of Article I of the Convention.

The Conference reaffirms the undertaking in Article I never in any circumstance to develop, produce, stockpile or otherwise acquire or retain weapons, equipment or means of delivery designed to use such agents or toxins for hostile purposes or in armed conflict, **in order to exclude completely and forever the possibility of their use.**" [Emphasis added]

The BTWC has no provisions for verification or for the monitoring of compliance. As already noted above, at the Third Review Conference in September 1991 following the Persian Gulf war of 1990/1991, the States Parties established an Ad Hoc Group of Governmental Experts to examine possible verification measures from a scientific and technical viewpoint. The final report of VEREX was considered by a Special Conference in September 1994 which established an Ad Hoc Group (AHG) to consider appropriate measures, including possible verification measures, and draft proposals to strengthen the Convention, to be included, as appropriate, in a legally binding instrument. By December 1999 this had met 17 times and had issued the tenth draft of the rolling text of the Protocol to strengthen the BTWC.

All the essential elements for the Protocol are now in the rolling text – Mandatory Declarations, Non-Challenge Visits (both focussed and random) and Compliance Concern Investigations together with measures to strengthen the implementation of Article X (cooperation for peaceful purposes) and other Articles of the BTWC. Although there is much proliferation of square brackets indicating alternatives, there is nothing new required to strengthen the BTWC. Recent political developments show that the political will is there to complete the negotiations and finalize the Protocol. The Ad Hoc Group has a real opportunity to do this during the sessions scheduled for 2000.

The importance of countering the possible use of chemical and biological materials for terrorist purposes has been recognized by the Heads of State and Government of the G7 who in their June 1996 summit in Lyon, France declared that:

> "Special attention should be paid to the threat of utilization of nuclear, biological and chemical materials, as well as toxic substances, for terrorist purposes." [40]

Subsequent action included the negotiation in the Sixth Committee of the General Assembly of the International Convention for the Suppression of Terrorist Bombings which was adopted by the General Assembly in Resolution A/RES/52/164 on 15 December 1997. This Convention makes it an offence if a person "unlawfully and intentionally delivers, places, discharges or detonates an explosive or other lethal device...". [41] The definition of "explosive or other lethal device" includes

> "A weapon or device that is designed, or has the capability, to cause death, serious bodily injury or substantial material damage through the release, dissemination or impact of toxic chemicals, biological agents or toxins or similar substances or radiation or radioactive material."

5.2 CONTROLS

All States Parties to the BTWC have an obligation under Article III which states that:

> "Each State Party to this Convention undertakes not to transfer to any recipient whatsoever, directly or indirectly, and not in any way to assist, encourage, or induce any State, group of States or international organizations to manufacture or otherwise acquire any of the agents, toxins, weapons, equipment or means of delivery specified in article I of the Convention."

Consequently, they are obligated to implement appropriate national control measures to meet this requirement. It is becoming apparent that an increasing number of states around the world are introducing export control regimes in order to prevent exports being misused for prohibited purposes. In addition, the import-export mechanism for Iraq set up by the United Nations Security Council requires that all states exporting particular materials or equipment to Iraq notify these to an UNSCOM import/export monitoring unit located in New York.[42] States are expected to have established appropriate national mechanisms to meet this requirement.

A number of states have formed an informal ad hoc grouping, known as the Australia Group, with the aim of harmonizing controls of both materials and dual purpose equipment that might be misused for chemical or biological programs in contravention of the undertakings in the CWC and the BTWC. Some 30 states are now members of the Australia Group. Although some members of

the Non-Aligned Movement (NAM) claim that the AG controls are discriminatory, there is little if any evidence to support this. [43] Indeed, it is becoming
increasingly apparent that we are living in a world in which more and more
states, both developed and developing, are concerned about the possible dangers
from pathogenic or modified microorganisms to public health and environmental safety. [44] This has led to the negotiation of legally binding protocols
for Advanced Informed Agreement (the Biosafety Protocol) for transfers of
living modified organisms; this negotiation was completed on 29 January 2000
with the adoption by the Conference of the Parties to the Convention on Biological Diversity of the Cartagena Protocol on Biosafety. [45] As of 15 September 2000, 75 States had signed the Protocol.

In addition, the concern about the possible acquisition of such materials by
terrorists has led to a strengthening of national controls. For example, the
United States on 15 April 1997 introduced new rules requiring the registration
and inspection of all facilities holding, using or transferring "select agents" – a
list of biological agents and toxins that closely parallel the agents generally
regarded as candidates for use as biological weapons.

There are encouraging signs that dangerous chemicals and pathogens will
increasingly be subject to national, regional and international controls to ensure
public health and environmental safety.[46] Such a framework of controls on
the handling, use and transfer of such materials will, over time, build transparency and confidence that these materials are not being misused and thereby
contribute to a stronger CBW security regime.

5.3 PROTECTIVE MEASURES

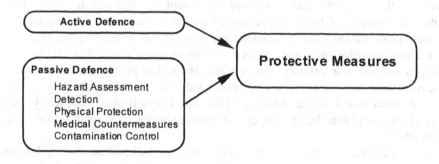

Figure 3. Protective Measures:

These comprise both active defense such as interception and destruction of incoming missiles and passive defense made up of hazard assessment, detection
and identification, physical protection and medical countermeasures together

with contamination control. An integrated set of broad band defensive measures are needed to help to reduce the effectiveness of biological weapons.

The importance of protective measures has been reinforced recently by the work of the NATO Senior Defence Group on Proliferation (DGP).[47] The three year study carried out by the DGP led to the recognition that "NBC weapons are quite different from one another as are their characteristics and their potential military effect." [48] In other words, biological weapons must be regarded as different from chemical weapons and from nuclear weapons. This consideration led the DGP to "draw particular attention to the importance of protection for deployed forces, given NATO's new roles and missions and the regional nature of the risk." The DGP study has resulted in the definition of a series of overarching principles to guide NATO's defense response to proliferation. These include:

"a. Maintain freedom of action and demonstrate to any potential adversary that the Alliance will not be coerced by the threat or use of NBC weapons.
b. Reassure both Allies and coalition partners of the Alliance's ability effectively to respond to, or protect against, NBC threats or attacks.
c. Complement non-proliferation efforts with a mix of military capabilities that devalue NBC weapons, by reducing the incentives for, and raising the cost of acquisition.
d. Emphasize system mobility, given that NBC proliferation risks are expected to be primarily regional in character and that NATO forces may be called upon to operate beyond NATO's borders." [49]

A number of priorities are set out: to counter proliferation risks, NATO has greatest need for capabilities for the detection (both point and standoff) of biological (and chemical) agents, as well as attack identification and warning, together with NBC individual protective equipment for deployed forces. In June 1997, the meeting of the North Atlantic Council reaffirmed its commitment to address proliferation risks as an integral part of the Alliance's ongoing response to the new security environment. Its final communiqué stated that "While prevention remains our primary aim, we recognize that proliferation of nuclear, chemical and biological (NBC) weapons and their means of delivery ... can pose a direct threat to the Alliance." [50] The Council also specifically welcomed the emphasis being put on "improving protection against biological weapons."

A similar recognition of the importance of chemical and biological defense has been shown in the United States by the Quadrennial Defense Review which stated that "In particular, the threat or use of chemical or biological weapons (CBW) is a likely condition of future warfare, including the early stages of war to disrupt U.S. operations and logistics."[51] It went on to argue that "Moreover, given that the United States will most likely conduct future operations in coalition with others, we must encourage our friends and allies to train and equip their forces for effective operations in CBW environments." The importance of addressing the protective measures in coalitions facing a chemical or biological weapons possessor has recently been discussed.[52] It is clear that action is needed to improve these capabilities – and that particular priority needs to continue to be placed on biological defense.

A recent perceptive article on NBC defense in the 21st century rightly concluded that "protection will continue to be needed through the distant future, and not only for the military, but also for the police and the medical and civil defence, and for the general public." [53]

Advances in biotechnology have a great deal to offer in terms of achieving effective protective measures. The principal requirement is for rapid, real-time detection and identification of a biological agent attack in order for physical protective measures such as respirators or collective protection to be donned. Biotechnology offers several techniques which can achieve rapid and accurate detection followed by identification of the agent. The second area of biological defense in which biotechnology has much to offer is that of medical counter-measures where new and improved vaccines can be developed and produced. A further area in which biotechnology has a role to play is in the analysis of samples. Such analysis needs to be carried out rigorously, particularly if the results are to be used to underpin the submission of evidence nationally or internationally, such as to a future BTWC Organization or to the United Nations, to demonstrate that a biological weapons attack has indeed occurred. Indeed, it is evident that in all of these areas, biotechnology is the key, and indeed is frequently the **only** realistic route, to providing effective protective measures.

5.4 DETERMINED RESPONSES

The importance of a determined response to non-compliance with the BTWC or the CWC has to be emphasized. There is little doubt that the lack of response by the international community to the proven use of chemical weapons during the Iraq-Iran war of the 1980s sent entirely the wrong message to the leaders in the region – that chemical weapons were apparently an acceptable form of warfare and their use would not incur significant penalties. The united stand taken against Iraq following the invasion of Kuwait in August 1990 and the strong UN Security Council resolution 687 sent the clear message that the possession of weapons of mass destruction is unacceptable. However, the events of the first two months of 1998 when Iraq once again failed to comply with the Security Council resolutions, demonstrated a disappointing lack of unanimity in the Security Council about the need to take determined action to force Iraq to comply. Although subsequent Security Council action under resolution S/RES/1284 on 17 December 1999 [54] replaced UNSCOM by a new subsidiary body of the Council, the United Nations Monitoring, Verification and Inspection Commission (UNMOVIC), there has been no indication by October 2000 that Iraq will accept this Commission or that the Security Council is prepared to take steps to enforce this resolution. There is a real danger that if a firm line is not taken with

Iraq, other would-be possessors of CBW may judge that such weapons can be obtained with impunity with the disastrous consequences that would follow for international peace and security.

It needs to be recognized that determined responses to non-compliance are necessary to underpin the other elements of the web of deterrence. There is little point in establishing strong rules if these can be flouted with impunity. The consequences for the global economy and stability will be costly should states judge that the norm can be broken without incurring an unacceptable penalty. States need to recognize that deliberate and determined responses to non-compliance ranging from diplomatic approaches through sanctions to the threat and ultimately the use of armed intervention is a vital element of the web of deterrence.

6. Prospects

The web of deterrence – comprising comprehensive prohibition, broad monitoring and controls, broad band protective measures, both active and passive, and determined national and international responses to non-compliance – will continue to be of vital importance into the foreseeable future. There is a clear need for an early completion of an effective BTWC Protocol as biological weapons currently present the greatest danger of all weapons of mass destruction. Integrated protective measures against biological weapons will significantly degrade the attractiveness of such weapons and biotechnology offers immense potential benefits, especially in rapid detection and improved medical countermeasures. The global moves to improve public health and environmental safety will help to strengthen the national and regional framework within which materials that are potentially harmful, such as toxic chemicals and pathogens, are handled, used and transferred. This will contribute over time to increasing transparency and confidence that such materials are being used for permitted purposes. There is significant complimentarity and synergy between the security and the public health and environmental safety initiatives which should together help to ensure that the 21st century is a safer and more secure world for all.

7. References

1. Pearson, G.S. (1998) The Threat of Deliberate Disease in the 21st Century, in Amy E. Smithson (ed), *Biological Weapons Proliferation: Reasons for Concern, Courses of Action*, The Henry L. Stimson Center, Report No 24, January.

2. United Nations (1971) Convention on the Prohibition of the Development, Production and Stockpiling of Bacteriological (Biological) and Toxin Weapons and on their Destruction, General Assembly resolution 2826(XXVI), 16 December. Also available as HMSO, Cmnd 5053, 1972.

3. United States Congress, Office of Technology Assessment, (1993) Proliferation of Weapons of Mass Destruction: Assessing the Risks, OTA-ISC-559, S/N 052-003-01335-5, dated 5 August; United States Congress, Office of Technology Assessment (1993) Background Paper, Technologies Underlying Weapons of Mass Destruction, OTA-BP-ISC-115, S/N 052-003-01361-4, dated December.

4. Dando, M.R. (1999) *Biotechnology, Weapons and Humanity*, British Medical Association, Harwood Academic Publishers, The Netherlands.

5. United Nations Security Council Resolution 687 (1991) Security Council Resolution establishing detailed measures for a cease fire-in between cease fire, including deployment of a United Nations Observer Unit; arrangements for demarcating the Iraq-Kuwait borders; the removal or destruction of Iraqi weapons of mass destruction and measures to prevent their reconstitution, under the supervision of a Special Commission and Director General of the IAEA; and creation of a compensation fund to cover direct loss and damage resulting from Iraq's invasion of Kuwait, S/RES/687(1991), 3 April.

6. United Nations (1992) *The Third Review Conference of the States Parties to the Convention on the Prohibition of the Development, Production and Stockpiling of Bacteriological (Biological) and Toxin Weapons and on their Destruction,* Geneva, 9–27 September 1991, BWC/CONF.III/23, Geneva.

7. United Nations (1993) *Ad Hoc Group of Governmental Experts to Identify and Examine Potential Verification Measures from a Scientific and Technical Standpoint, Report* BWC/CONF.III/VEREX/9, Geneva.

8. United Nations (1994) Special Conference of the States Parties to the Convention on the Prohibition of the Development, Production and Stockpiling of Bacteriological (Biological) and Toxin Weapons and on their Destruction, Final Report, BWC/SPCONF/1 Geneva, 19–30 September.

9. United Nations (1998) Procedural Report of the Ad Hoc Group of the States Parties to the Convention on the prohibition of the Development, Production and Stockpiling of Bacteriological (Biological) and Toxin Weapons and on their Destruction, BWC/AD HOC GROUP/43, 15 October.

10. David Davis, UK Minister of State for Foreign & Commonwealth Affairs (1996) Statement to the Fourth Review Conference of the Biological and Toxin Weapons Convention, Geneva, 26 November.

11. United States Arms Control and Disarmament Agency, 1997 Annual Report, available at http://www.acda.gov/reports/annual/anrpt_97.htm

12. Kaplan, D.E. and Marshall, A. (1996) *The Cult at the End of the World:*

28

The Incredible Story of Aum, Hutchinson, London; United States Senate, Committee on Governmental Affairs (1996) *Staff Statement, Global Proliferation of Weapons of Mass Destruction: A Case Study on the Aum Shinrikyo,* 31 October and 1 November 1995, S. Hg. 104-422, p.47 et seq., US Government Printing Office, Washington, DC.

13   Pearson, G.S. (1999) *The UNSCOM Saga: Chemical and Biological Weapons Non-Proliferation,* Macmillan/St. Martin's Press.

14.   United Nations (1995) Eighth Report of the Secretary-General on the status of implementation of the plan for the ongoing monitoring and verification of Iraq's compliance with relevant parts of Section C of Security Council resolution 687 (1991), S/1995/864, 11 October.

15.   United Nations Security Council (1992) Note by the President of the Security Council, S/23500 dated 31 January.

16.   NATO (1994) Declaration of the Heads of State and Government participating in the meeting of the North Atlantic Council held at NATO Headquarters, Brussels, on 10-11 January 1994, Press Communiqué M-1(94)3, 11 January.

17.   NATO (1994) Alliance Policy Framework on Proliferation of Weapons Of Mass Destruction, issued at the Ministerial Meeting of the North Atlantic Council held in Istanbul, Turkey on 9 June 1994, Press Release M-NAC-1(94)45, 9 June.

18.   NATO (1994) Communiqué issued by the Ministerial Meeting of the North Atlantic Council, NATO Headquarters, Brussels, 1 December 1994, published in NATO Review, No. 6 December 1994 - No. 1 January 1995, pp. 26-8.

19.   NATO (1995) Ministerial Meeting of the North Atlantic Council in Noordwijk aan Zee, The Netherlands, 30 May 1995, Final Communiqué, Press Communiqué M-NAC-1 (95)48, 30 May.

20.   NATO (1997) Final Communiqué, Meeting of the North Atlantic Council in Defence Ministers Session, Press Release M-NAC-D-1(97)71, Brussels, 12 June.

21.   The White House, Office of the Press Secretary (1998) State of the Union Adress by the President, 27 January 1998. Available at http://library.whitehouse.gov/

22.   United Nations (1998) "Working Paper submitted by the United Kingdom of Great Britain and Northern Ireland on behalf of the European Union", BWC/AD HOC GROUP/WP. 272, 9 March.

23.   Australian Permanent Mission (1998) Address by the Permanent Representative of Australia to the Conference of Disarmament, His Excellency Mr John B Campbell to the BWC Ad Hoc Group, Geneva, 9 March.

24.   Mision Permanente de Colombia (1998) Statement by the Non-Aligned Movement and other countries, Geneva, 13 March.

25.   G8 Foreign Ministers (1998) Conclusions of G8 Foreign Ministers: 9 May. Available at http://birmingham.g8summit.gov.uk/

26. Non-Aligned Movement (1998) Communiqué, Ministerial Meeting of the Coordinating Bureau of the Non-Aligned Movement, Cartagena des Indias, Colombia, 19-20 May, paragraph 107. Available at http:// www. nam.gov. Za/cartagena98.html

27. Non-Aligned Movement (1998) Communiqué, Ministerial Meeting of the Coordinating Bureau of the Non-Aligned Movement, Cartagena des Indias, Colombia, 19-20 May, paragraph 107. Available at http://www.nam.gov. Za/cartagena98.html

28. Office of the Press Secretary (1998) Remarks by the President at the United States Naval Academy Commencement, The White House, 22 May. Available at http://www.pub.whitehouse.gov

29. Tony Lloyd (1998) Time to Accept the Realities of the Control of Biological Weapons, Address to the Ad Hoc Group, Geneva, 22 June.

30. United States – China Joint Presidential Statement (1998) Joint Statement on Biological Weapons Convention, Beijing, 27 June. Available on http:// www.usia.gov/regional/ea/uschina/bioweps.htm

31. United Nations (1998) Ad Hoc Group of the States Parties to the BTWC, Working Paper, BWC/AD HOC GROUP/WP.296, 10 July.

32. United Nations (1998) Working Paper submitted by Australia, Declaration of the Informal Ministerial Working Meeting on the Negotiation towards Conclusion of the Protocol to Strengthen the Biological Weapons Convention, BWC/AD HOC GROUP/WP. 324, 9 October.

33. NATO (1999) Washington Summit Communiqué, issued by the Heads of State and Government participating in the meeting of the North Atlantic Council in Washington D.C. on 24th April 1999. An Alliance for the 21st Century. Press Release NAC-S(99)64, 24 April. Available at http://www. nato.int/docu/pr/1999/p99-064e.htm

34. European Union (1999) Common Position of 17 May 1999 adopted by the Council on the basis of Article 15 of the Treaty on European Union, relating to progress towards a legally binding Protocol to strengthen compliance with the Biological and Toxin Weapons Convention (BTWC), and with a view to the successful completion of substantive work in the Ad Hoc group by the end of 1999, (1999/346/CFSP), official Journal of the European Communities, L 133/3-4, Vol. 42, 28 May. Available at http://europa.eu. int/eur-lex/en/oj/1999/1_1331990528en.html

35. NATO (1999) Final Communiqué, Ministerial Meeting of the North Atlantic Council held at NATO Headquarters, Brussels on 15 December 1999, Press Release M-NAC2-(99)166, 15 December. Available at http:// www.nato.int/docu/pr/1999/p99-166e.htm

36. Pearson, G.S (1993) Prospects for Chemical and Biological Arms Control: The Web of Deterrence, *The Washington Quarterly*, Spring, 145–162.

37. Pearson, G.S. (1998) The Vital Importance of the Web of Deterrence, Proceedings, *Sixth International Symposium on Protection against Chemical and Biological Warfare Agents*, Stockholm, 10-15 May, pp. 23-31.

38. United Nations (1997) List of States Parties to the Convention on the Prohibition of the development, Production and Stockpiling of Bacteriological (Biological) and Toxin Weapons and on their Destruction, BWC/AD HOC GROUP/INF.11, 8 September.

39. United Nations (1996) Fourth Review Conference of the Parties to the Convention on the Prohibition of the Development, Production and Stockpiling of Bacteriological (Biological) and Toxin Weapons and on their Destruction, Final Declaration, Final Document, BWC/CONF.IV/9, 6 December.

40. United Nations (1996) Letter dated 5 July 1996 from the Permanent Representative of France to the United Nations addressed to the Secretary-General, A/51/208, S/1996/543, 12 July.

41. United Nations General Assembly (1997) Measures to eliminate international terrorism, Report of the Sixth Committee, A/52/653, 25 November.

42. Pearson, G.S. (1998) Article III: Further Building Blocks, University of Bradford Briefing Paper No. 13, October. Available at http://www.brad.ac.uk/acad/sbtwc

43. Roberts, B. (1996) Article III: Non-Transfer, in Graham S. Pearson and Malcolm R Dando (eds), *Strengthening the Biological Weapons Convention: Key Points for the Fourth Review Conference*, Quaker United Nations Office, Geneva, 23 September.

44. Pearson, G.S. (1997) The Complimentary Role of Environmental and Security Biological Control Regimes in the 21st Century, *JAMA* **278(5),** 369-372.

45. Cartagena Protocol on Biosafety. Available at http://www.biodiv.org/biosafe/protocol

46. Pearson, G.S. (1998) Article X: Some Building Blocks, University of Bradford, Department of Peace Studies, Briefing Paper No. 6, March; Pearson, G.S. (1998) Article X: Further Building Blocks, University of Bradford, Department of Peace Studies, Briefing Paper No. 7, March. Available at http://www.brad.ac.uk/ acad/sbtwc

47. The NATO Senior Defense Group on Proliferation (DGP) was established following the decision of the NATO Heads of State and Government meeting of the North Atlantic Council held at NATO Headquarters, Brussels, on 10-11 January 1994 to intensify and expand NATO's political and defense efforts against proliferation (see Press Communiqué M-1(94)3, 11 January 1994). Six months later, NATO issued the Alliance Policy Framework on Proliferation of Weapons Of Mass Destruction at the Ministerial Meeting of the North Atlantic Council held in Istanbul, Turkey on 9 June 1994 (see Press Release M-NAC-1(94)45, 9 June 1994). This framework set out various aspects of the defense dimension stating that "As a defensive Alliance, NATO must therefore address the military capabilities needed to discourage WMD proliferation and use, and if necessary, to protect NATO territory, populations and forces".

48. NATO (1995) NATO's Response to Proliferation of Weapons of Mass Destruction: Facts and the Way Ahead, NATO Press Release (95)124, 29 November. See also Carter, A.B. and Omand, D.B. (1996) Countering the Proliferation Risks: Adapting the Alliance to the New Security Environment, *NATO Review*, September, 10-15.

49. Carter, A.B. and Omand, D.B. (1996) Countering the Proliferation Risks: Adapting the Alliance to the New Security Environment, *NATO Review*, September, 10-15. Available at http://www.natio.int/docu/review/articles/9605-3.htm

50. NATO (1997) Final Communiqué, Meeting of the North Atlantic Council in Defence Ministers Session, Press Release M-NAC-D-1(97)71, Brussels, 12 June.

51. US Department of Defense (1997) William S. Cohen, Secretary of Defense, Report of the Quadrennial Defense Review, May.

52. Roberts, B. and Pearson, G.S. (1998) Bursting the biological bubble: how prepared are we for biowar?, *Jane's International Defence Review* **31(4).**

53. Bovallius, A. (1997) NBC in the 21st Century: A widened threat spectrum, ASA Newsletter 97-6, 4 December, issue no 63, 1, 3.

54. United Nations Security Council (1999) Resolution 1284 (1999) adopted by the Security Council at its 4084th meeting on 17 December 1999, S/RES/1284 (1999), 17 December. Available at http://www.un.org/unscom/ Keyresolutions/sres99-1284.htm

LESSONS OF THE CHEMICAL WEAPONS CONVENTION FOR THE BTWC PROTOCOL

ALEXANDER KELLE
Peace Research Institute Frankfurt
Leimenrode 29
D–60322 Frankfurt
Germany

JIRI MATOUSEK
Institute of Environmental Chemistry and Technology
Faculty of Chemistry, Brno University of Technology
Purkynova 118
CZ–612 00 Brno
Czech Republic

1. Commonalities and Differences of CBW Agents

Chemical warfare agents on the one hand and biological and toxin warfare agents on the other enjoy a number of commonalities. This basic statement lies at the heart of the assumption that there are lessons to be learned from the structure, content and implementation experience of the Chemical Weapons Convention (CWC) for the negotiations of the Compliance Protocol to the Biological and Toxin Weapons Convention (BTWC or BWC). In our discussion of these commonalities we follow the subdivision of issues which was put forward by Robinson [1]. Thus, in the remainder of this first section we will take up the common taboo against both CW and BW, the issue of toxins as the overlapping category of agent covered by both CWC and BWC, the problems of dual-use and of technological change, and the question of effective protection against CBW.

This notion of commonalities among CW and BW is also supported by the historical coverage of CBW weapons in arms control or other international agreements. First efforts in this regard date back to 1868, the year in which the Declaration of St. Petersburg was formulated. Motivated by a mainly humanitarian impetus, it stated that the employment of arms which needlessly aggravate the suffering of disabled soldiers, or render their death inevitable, would be contrary to the laws of humanity. The international document which made a

A. Kelle et al. (eds.), The Role of Biotechnology in Countering BTW Agents, 33–45.
© 2001 Kluwer Academic Publishers. Printed in the Netherlands.

first explicit reference to CW-like weaponry is the Declaration on Laws and Methods of Conducting Wars – signed in Brussels, 27 August 1874, but never entered into force – prohibiting *inter alia* the employment of poisonous and poisoned weapons. This prohibition then was included in the Convention on Laws and Customs of Land Warfare adopted at the 1st International Peace Conference in The Hague in July 1899.

In the 1925 Geneva Protocol CW and BW were – following a Polish initiative – for the first time treated as different weapon-categories, but on the same footing. The protocol prohibits the use of asphyxating, poisonous or other gases and of all analogous liquids, materials and devices, *as well as* the use of bacteriological methods of warfare. Unfortunately, the non-use norm inscribed in this legal document has been weakened by unilateral reservations made by a number of states. These reservations or conditions limit the Protocol's applicability to states parties to it and/or to first use only. [2]

Yet, there are a number of differences among the CW and BW agents, which not the least have informed the separated treatment of the two in the late 1960s, resulting in the conclusion of the BWC in 1972 with all its shortcomings that have come to the fore since its entry into force. The differences between the two categories will also be briefly taken up in the following five subsections. A more detailed assessment of how far the analogy between CW and BW carries will then be undertaken in the main section of this chapter, covering four categories of compliance measures: declarations states parties will have to submit to a future BWC organization, challenge inspections, investigations of alleged use, and visits.

1.1 THE TABOO AGAINST CBW

The aversion against "poison weapons" – be they CW or BW agents – can be traced back two and a half thousand years. The Manu laws of India are cited as are stipulations in the Koran, which served as warfare regulations for the Saracens. This leads Robinson to the conclusion that what we see as the underlying principle of both the CW and BW control regimes:

> "is surely the operation of a taboo, an ancient norm of behaviour that invests resort to poisons and germs as weapons of war with an opprobrium which may expose transgressors to serious sanction, a taboo that stretches far back through the centuries and across cultures." [1]

Despite the widespread use of CW during World War I, the taboo can be said to have been reinvigorated over the course of the 20th century. Especially, the non-use of CW during World War II both reflects the taboo and was essential for its further strengthening. Generally three explanatory factors are given for CW non-use during World War II: first, reciprocal warnings of the belligerents of retaliation in kind as response to a first use of CW; second, the level of unpreparedness within the military establishments to use this type of weapons and;

third, "a general feeling of abhorrence on the part of governments for the use of CB weapons". [3] Although it has been argued that the uncertain military value of CW, the resistance of military cultures bound to traditional ways of war-fighting, as well as the additional logistical efforts necessary to integrate CW into military doctrines can by and large account for the resistance of military establishments to prepare for employing CW during World War II [4], it has been shown that these motivations are insufficient to account for the CW non-use in the absence of the underlying taboo. In other words, "the existence of a stigma against using CW was a necessary condition for the nonuse of CW." [5]

In the period since World War II the taboo has continuously and gradually been strengthened. Despite US use of CW during the Vietnam War and repeated CW use in the Iran/Iraq war the conviction that chemical warfare is completely inacceptable among civilized nations has been upheld and reinforced. [6]

1.2 TOXINS

Toxins represent the area of overlap between CW and BW. Under "normal" circumstances produced by living organisms, they fall into the purview of the BWC. Yet, not being living matter, they perfectly fit the CWC's definition of toxic chemicals. Hence, they are covered in both the CWC and the BWC. Article I of the BWC includes in the scope of the convention "toxins whatever their origin or method of production". In the CWC toxins are not only covered by Article II 2., two of them, i.e. saxitoxin, and ricin, have been included in Schedule 1 of the CWC's *Annex on Chemicals*.

With the CWC in force only for a short period of time and the BWC protocol still under negotiation, the jury is still out on the question posed by Robinson, whether it is "prudent to have two different, and therefore potentially divergent, control regimes for toxins". [1] A tentative answer to this question would point to the comprehensive character of the prohibition of toxins under both CWC and BWC, which should in principle allow for the formulation of more concrete rules and procedures which allow the implementation without creating any loopholes. However, the two conventions being set up in the way that they are, it would seem more advisable to live with some organizational duplication, because of the overlap between CWC and BWC, rather than to untangle the basic fabric of either of the two conventions in order to eliminate the overlap. Should such duplication of work occur once the Organization for the Prohibition of Biological Weapons (OPBW) will have been set up, it will be a relatively modest price to pay for the comprehensive coverage of toxins. The final answer on the question of duplicating work, however, will have to assess the control regime for toxins established under the BWC protocol *and* its implementation through OPBW and states parties.

1.3 DUAL-USE AND TECHNOLOGICAL DEVELOPMENTS

Many toxic chemicals cannot be used for weapons purposes only, but have perfectly legitimate civil applications. The same can be said for equipment, material, and know-how required for the production of CW agents. As Robinson reminds us, the origins of modern chemical warfare indeed go back to the advent of chemical industry during the first decades of the 20th century: "mass destruction CW originated in the high-technology chemical industry of Europe at the time of the First World War, pressed forward by the desire of academic chemists to contribute their skills [...] to the war efforts of their countries." [1, 7] What was high-technology chemical knowledge and processes back then, however, has become mostly low-technology today and is thus even more widely accessible, thereby aggravating the dual-use problem with regard to CW.

With regard to biological weapons the dual-use problem presents itself in a slightly different manner with respect to two issues. First, all biological warfare agents are occurring naturally, outside any military context. This is true for only a fraction of chemical warfare agents. Those CW agents listed on the CWC's Schedule 1 for example have few civilian applications. In other words: while in the CW context, the dual use problem by and large runs up to the level of CW precursors and touches only some of the agents themselves, all BW agents are dual use, in the sense that they are the causative agents of naturally occurring diseases – endemic in certain regions of the world – that can be misused for weapons purposes.

Second, in contrast to CW, the potentially most dangerous BW-related technology is not low-tech World War I style, but derives from the potential for misuse of state of the art modern biotechnology and genetic engineering. In a 1993 survey of the field, Bartfai, Lundin and Rybeck identified three types of potential misuse: "the enhancement of bacterial and viral virulence, heterologous gene expression and protein engineering of toxins, and genetic weapons." [8] As Dando has pointed out in a more recent study with respect to the possibility of the latter being realized:

> "the gap between theoretical possibility and practical reality seems to be closing steadily. Significant new successes in the application of gene therapy, in particular, should be carefully monitored an assessed. While it can be hoped that ethnically specific weapons will never become a reality, it would be foolish to imagine that they are an impossibility or that incredibly precise targeting might not become possible." [9]

Thus, the dual-use challenge in the BW field presents itself in a similar but not identical way to the one in the CW area. The differences between the two manifestations of the dual-use problem – which to a considerable extent are caused by the differences in technology levels, i.e. high tech in the BW field and mostly low tech in the CW area - should also be expected to necessitate different approaches in dealing with them. In terms of negotiating and later implementing the BWC protocol, the dual-use problem is bound to impact on a number of issue areas, including compliance measures.

1.4 PROTECTION AGAINST CBW

Unlike in the case of an attack with nuclear weapons, defenses against the use of CBW are still possible after the fact, i.e. after an attack has occurred. This is so because – as Robinson points out – the most efficient dispersal systems for CBW utilize the "atmospheric environment" as "an integral part of the overall weapon system." [1] Yet, an essential prerequisite for such a "late" defense is a detection capability to recognize an attack in the first place. It requires furthermore sufficient protective equipment as well as the training to use it properly. This may sound pretty obvious, but it has wide ranging implications for both procurement and training activities that have to occur well before such an attack takes place.

In addition, when considering the prospects and options for defenses against the use of CBW a distinction between the two categories of WMD is in order. This concerns first and foremost the characteristics of CW or BW agents, the amounts needed to conduct a large scale attack – which can be conceived of as the worst case and will therefore serve as the basis for the following reasoning – and resulting from it, the most promising defensive approach. According to Utgoff:

> "Offsetting the possibility that destroying BW attack capabilities before they could reach their targets may be far more difficult than in the case of CW attack capabilities, preventing BW agents from damaging populations whose neighborhoods they reach is significantly easier." [10]

In other words, active defenses are more likely to be effective against CW attacks, while for the protection against BW attacks passive defenses might be more promising. Why is this so? In the case of CW, active defenses, i.e. the interception of a delivery vehicle carrying CW before this reaches its target, are more promising than in case of a BW attack simply because the amount of CW agents needed to inflict large-scale damage is much higher. Consequently, any aggressor would have to use large numbers of missiles or aircraft – if this indeed were the preferred means of delivery, instead of a ship or truck – in order to deliver the agents to their targets. This, in turn, would be relatively easy to detect and enable the targeted state, or a security guarantor, to respond either in a preemptive mode or during the flight time of the delivery vehicles.

Of course, active defenses against such attacks cannot be expected to prevent any CW agents from reaching their targets. Especially if one takes into account the possibility of an attacker who uses ballistic missiles to divide the warhead into submunitions, which increase the demands on the active defenses of the attacked state considerably. But even then, the attacked armed forces, but also civilian populations, stand a good chance to protect themselves effectively against the attack by means of passive defense simply by wearing masks. In addition, protective gear can be made available to larger portions of the armed forces, while civilians most likely would have to remain indoors, preferably in sealed rooms. For these measures to be implemented, warning systems with highly sensitive sensors showing the presence of CW agents will have to be installed and operating in the first place. These sensors will also enable the de-alerting and cessation of the protective measures, e.g. after a decontamination has taken place.

In contrast to CW, active defenses against attacks with BW are much more difficult to employ, because of the small amount of BW agents required to achieve a strategic level of damage. Only 100 kg of anthrax spores effectively dispersed may lead to up to 3 million dead. For the delivery of that amount no large scale attack with aircraft or ballistic missiles is required – consequently, no such attacks can be intercepted by means of active defenses. Nevertheless, passive defenses stand a good chance to protect both military and civilians against a BW attack. Five characteristics of BW agents account for this:
- first, besides aerosol dispersion, the most likely route of BW agents to attack successfully is through the digestive system. Given the limited resistance of most BW to high temperatures, eating well cooked food and drinking boiled water provides a sufficient defense for this scenario.
- second, practically none of the BW agents, that attack through the skin, are usable for large-scale attack. The few exceptions that exist to this rule can be countered by wearing protective cloths or simple washing of the skin.
- third, for most of the traditional BW agents, vaccines and antidotes are available and are improved continuously.
- fourth, biotechnology can be expected to increase both productivity in the manufacture of vaccines and to decrease the response time necessary to develop new ones against new BW agents.
- fifth, protective masks suitable for defenses against CW usually serve well against BW agents. In some cases (of BW agents) even high efficiency dust masks are sufficient. [10]

All these protective measures, of course, depend on an appropriate hazard assessment as well as the timely detection and identification of BW agents involved in an attack. [11, 12, 13]

2. CWC Compliance Measures and their Relevance for Strengthening the BWC

As the introductory section makes clear, the commonalities and differences between CW and BW will allow for the transfer of some provisions from the CWC to the BWC protocol context, will require the adaptation of other CWC provisions and prevent still another set of CWC measures from being transferred to the work of the Ad Hoc Group.

In the following subsections we will first present the relevant CWC provisions very briefly, then check the BWC content for corresponding provisions, give a brief assessment on the transferability of the CWC stipulations and, lastly offer an analysis of the AHG's "performance" with respect to its ability to produce consensual text for the issue under consideration. The cut-off date for consideration of the AHG's work is end-1999.

Given the multitude of issues, the AHG has to deal with – and in light of the preceding chapter by Graham Pearson – we will limit our discussion to the core issue area of the protocol, i.e. compliance measures, including declarations, routine inspections/visits, challenge inspections, and investigations of alleged use. Contrary to the usage of the term in the CWC, in the context of the BWC protocol inspections only refer to what might be considered an equivalent to CWC challenge inspections. The concept of routine inspections as found in the CWC is not being dealt with in this area of the BWC protocol negotiations. Rather, an equivalent of routine inspections might eventually be found in the context of visits, but even here some delegations in the AHG – like the US and some NAM countries – are very hesitant to agree on anything that looks like routine on-site measures, whatever the terminology attached to them might be.

2.1 DECLARATIONS

The CWC requires declarations from its states parties on a number of issues, most importantly in Article III on the possession of chemical weapons, old and abandoned chemical weapons, and chemical weapons production facilities. Given the obligation under the BWC for all member states to destroy their stocks of BW within nine month after the convention's entry into force, a declaration requirement on BW possession should produce only nil declarations. Nevertheless, since a state party might have "forgotten" to destroy its BW stocks, a declarations requirement should be included in the BWC protocol. The same is valid for old – e.g. in the form of spores – and abandoned BW as well as (past) BW production facilities.

Yet, in principle the major declaration task for states parties to the protocol is bound to fall into the category of dual-use facilities, which have legitimate civilian purposes, but could in a number of days be converted into a BW production facility if the decision were made. In the CWC context, requirements for facility declarations vary according to the three Schedules – contained in the Annex on Chemicals to the CWC – on which the chemicals are listed, as well as the annual amounts of these chemicals. As Tucker points out:

> "Basing the CWC's verification regime for chemical industry on the types and quantities of chemicals produced, processed, or consumed at various plant sites makes sense because a finite number of chemicals possess both the high toxicity and physiochemical properties required for warfare use." [14]

There are no comparable provisions in the BWC and also the Confidence Building Measures (CBM) agreed upon during the Second and Third BWC Review Conference in 1986 and 1991, respectively do not provide any guidance on how to incorporate industry declarations in the BWC protocol. However, differences in the substance matter of CW and BW preclude a simple copying of the CWC stipulations into the BWC protocol. First, in the BW realm there is not a comparably small number of known warfare agents amenable to being put on a list. Second, the already larger number of potential BW agents might continue to grow. This could be caused both by newly emerging diseases [15] and advances in genetic engineering [7, 8]

During the deliberations of the Ad Hoc Group, two ideal types of dealing with the question of declaration trigger can be identified. One takes the level of biocontainment as the yardstick for submission of facility declarations. Biosafety Level (BL)-4 has proven uncontroversial as trigger for declarations; BL-3, in contrast, is hotly contested as a stand-alone trigger. [16] Consequently, the second approach, i.e. to rely on a combination of activities or programs for the identification of suitable declaration triggers, will have to be taken into consideration for inclusion in a protocol text. While it seems likely that past BW programs, BW defense activities and vaccine production plants will have to be declared, it is far from clear how detailed the declarations will be or which declaration formats will be used. As evidenced by the Report of the 17th meeting of the AHG, even more work needs to be done on other conceivable declaration triggers or combinations thereof.[17]

If there is a single lesson for the BWC protocol to be drawn from CWC implementing experience so far, it is that states parties' failure to submit declarations in full and on time has to be sanctioned. Implementation of declaration-related CWC provisions showed some serious shortcomings. Even two years after the CWC's entry into force 29 out of more than 120 States Parties still had to submit their initial declarations. In addition, the declarations of a significant number of States Parties were incomplete or wrong, with the US amongst the "technically" non-compliant states.[18]

This experience then triggered the AHG to include during its July 1999 session a section on "Measures to Ensure Submission of Declarations" in Article III, Section D of the rolling text for the protocol. Although the exact scope and structure of these measures have still to be finalized, it seems agreed in the AHG that the:

> "provisions require the Director-General to report to each session of the Conference of the States Parties on the implementation of the declaration obligations and identify a number of punitive measures that might be applied should a state party not submit its initial or annual declarations within the (6) month period following the relevant deadline." [19]

2.2 CWC ROUTINE INSPECTIONS AND BWC PROTOCOL VISITS

Routine inspections under the CWC are based on declarations submitted by member states. They are:

> "periodic, pre-announced visits to declared government and commercial facilities to check the accuracy of declarations and to verify the absence of undeclared illicit production of chemical agents or the diversion of dual-use chemicals for military purposes." [14]

According to CWC Article VI each state party has to subject "toxic chemicals and their precursors listed in Schedules 1, 2 and 3 ... facilities related to such chemicals, and other facilities as specified in the Verification Annex to verification measures." The details of these measures are then spelled out in Parts VI to VIII of this Annex. Corresponding to the danger chemicals pose to the object and purpose of the CWC they are distributed across the three lists. For the most dangerous chemicals which are placed on Schedule 1 the CWC requests "systematic verification through on-site inspection and monitoring with on-site instruments". With the threat of chemicals on Schedules 2 and 3 decreasing, the less rigid are the respective CWC's verification provisions.

In principle, it is advisable to have a similar set of routine on-site measures included in the protocol to strengthen compliance with the BWC. Yet, at least three potential obstacles have to be taken into consideration when transferring the CWC's routine inspection. First, as in the context of the CWC's routine inspections are based on lists of chemicals, precautions have to be taken that when transferred into the BWC the scope of the BWC's Article I is not restricted. [20] Second, as outlined above, chemical warfare agents contained on the three CWC Schedules have few commercial applications. Potential BW agents, however, are endemic in certain regions of the world. Therefore, unlike in the case of CW agents, their presence does not necessarily provide a "smoking gun" indicating a treaty violation. Third, during the negotiations of the CWC chemical industries were rather supportive of the inspection regime as it developed. Concerning the ongoing negotiations in the AHG, however, pharmaceutical and biotech-industries are much more hesitant to accept – if not clearly opposed to, as in the case of US *PhRMA* – anything that looks like routine on-site measures in their facilities.

As a result of these obstacles, the gaps that still have to be bridged between AHG delegations' views concerning the role and character of visits in the future protocol are much wider. Proponents of non-challenge visits argue that such on-site measures increase the transparency of BW-relevant activities, enhance confidence in the treaty-abiding behavior of states parties by increasing the likelihood that violations are detected, contribute to the clarification of ambiguous declarations, and enhance cooperation among states parties. Critics of non-challenge visits counter that the detection of treaty violations through such weak on-site measures is highly unlikely. Rather, there exists – according to this group of states, which includes the US – the danger that visits lead to the loss of confidential business or national security information. Consequently, the frequency and intrusiveness of such on-site measures has to be minimized. [21] Towards the end of the 1999 AHG negotiating period:

> "differences in states' ideas for visits particularly focussed on the proposed provisions for clarification visits – i.e. visits to clarify declarations, and randomly-selected visits – i.e. infrequent mandatory visits to facilities selected on a random basis as a follow-up to declarations." [16]

2.3 FROM CHALLENGE INSPECTIONS TO FIELD AND FACILITY INVESTIGATIONS

Challenge inspections – in contrast to routine on-site measures – are exclusively concerned with suspicions of treaty violations, not the confirmation of treaty compliant behavior. CWC provisions that deal with such challenges are contained in CWC Article IX as well as Part X of the Verification Annex.

As in the case of the other categories of compliance measures, the BWC does not contain any relevant and practicable provisions. Although Article VI of the BWC stipulates that each "State Party to this Convention which finds that any other State Party is acting in breach of obligations deriving from the provisions of this Convention may lodge a complaint with the Security Council of the United Nations", this article has never been invoked. In addition to its awkward logic – not a suspicion, but a finding has to be presented by a state who wants the Security Council to take action [22] – there has, of course, always been the possibility that a veto of one of the permanent members of the Security Council would prevent any action following a complaint.

Yet, not only would it be desirable, but it is also judged as feasible by most AHG members to have a challenge inspection procedure included in the protocol which will be administered by the future BWC organization. As a result, the section on challenge inspections is – with respect to compliance measures – the one where most agreed upon language has been included in the rolling text.

One fundamental question which still has to be settled among AHG members relates to the initiation of an inspection and how an inspection that might be frivolous can be stopped. Basically, two different approaches can be envisaged. According to the first one, the so called "red-light" approach, an inspection request will go forward unless a majority – which would still have to be defined – of the executive council of the future organization decides otherwise. This position is supported by states – like the UK – who wish to see an effective compliance regime be set up. The "green-light" approach, in contrast, would allow for an inspection to proceed only if a majority of EC members approves it. This is championed by those states in the AHG who want to raise the threshold for triggering challenge inspections and thereby want to prevent frivolous challenges or minimize the risk of industrial espionage.

A second issue still to be settled in regard to challenge inspections relates to the transition from a field to a facility investigation. While a field investigation provides access to an area in which it it "suspected that biological weapons had been used", the latter one serves to check on a "facility that is suspected of hosting an illicit biological weapons program". [16] A first problem which still has to be settled in the AHG, of course, relates to the ability of the future BWC organization to distinguish between suspected BW use and a natural outbreak of a disease. [23] Once a field investigation is under way, a specific facility within the area investigated might appear as source of a contamination. As a result, it might become necessary to change the type of investigation from field to facility. However, a number of NAM countries are concerned that the future organization might be instrumentalized to gain through a field investigation "quicker and easier access to sensitive facilities than a facility investigation" would do. [16]

3. Conclusion

This chapter set out to discuss some aspects concerning the transferability of CWC provisions and their implementation to the BWC protocol. Both the common taboo against use of CBW agents and the coverage of toxins in both the CW and the BW control regimes strongly suggest the possibility of transferring certain provisions from one regime to the other. At the same time the differences in the dual-use character in the two areas as well as the level of technological developments required for the production and use of CW or BW agents respectively and the technicalities involved in the protection against CBW agents ask for a cautious approach in the application of CWC lessons to the BWC protocol negotiation.

What is more, the application of lessons learned in the CW context and the transfer of certain provisions to the BWC protocol not only hinges on technical questions. Rather, the political and legal context for controlling CBW has to be taken into account, too. In this regard one has to keep in mind that the CWC was a new international disarmament treaty, while the BWC is already in force for 25 years. This may seem banal, but is not to the extent that a number of negotiating positions undoubtedly are informed by the very experience with BWC implementation up to now: along these lines it can for example be taken for granted that the 1979 anthrax incident in Sverdlovsk lies at the heart of concerns that a practicable procedure for transforming field investigations into facility investigations has to be included in the protocol.

Two overall goals when completing the protocol negotiations have to be kept in mind from a technological point of view. First, the scope of Article I of the BWC must not be limited by any list or declaration trigger or format. Second, and related to this, the protocol must be made a flexible instrument that can

44

easily be adopted to new technological developments. This latter point is even more urgently needed in the BW context than in the realm of CW – largely because of the fact that the modification of existing or even creation of new BW agents through genetic engineering is increasingly becoming an option as biotechnology develops further. Members to the BWC protocol must by all means be in a position to keep this instrument easily up to date so as to cope with future cutting edge technological developments.

4. References

1. Robinson, J.P. (1996) Some Lessons for the Biological Weapons Convention from Preparations to Implement the Chemical Weapons Convention. in O. Thränert (ed.), *Enhancing the Biological Weapons Convention*, Verlag J.H.W. Dietz Nachfolger, Bonn, pp. 86-113.
2. Bothe, M. (1973) *Das völkerrechtliche Verbot des Einsatzes chemischer und bakteriologischer Waffen: Kritische Würdigung und Dokumentation der Rechtsgrundlagen*, Beiträge zum ausländischen öffentlichen Recht und Völkerrecht, Vol. 59, Köln/Bonn, Heymann.
3. SIPRI (1971) *The Problem of Chemical and Biological Warfare, Vol.4, CB Disarmament Negotiations, 1920-1970*, Almqvist and Wiksell, Stockholm.
4. Legro, J. (1995) *Cooperation Under Fire*, Cornell University Press, Ithaca, N.Y.
5. Price, R. (1995) A genealogy of the chemical weapons taboo, *International Organization* **49 (1)**, 73-103.
6. Price, R., Tannenwald, N. (1996), Norms and Deterrence: The Nuclear and Chemical Weapons Taboo, in Peter J. Katzenstein (ed.), *The Culture of National Security. Norms and Identity in World Politics*, Columbia University Press, New York, pp. 114-152.
7. Martinez, Dieter (1995) *Vom Giftpfeil zum Chemiewaffenverbot. Zur Geschichte der chemischen Kampfmittel*, Harry Deutsch Verlag, Frankfurt.
8. Bartfai, T., Lundin, S.J., and Rybeck, B. (1993) Benefits and Threats of Developments in Biotechnology and Genetic Engineering, in *SIPRI Yearbook 1993: World Armaments and Disarmament*, Oxford University Press, Oxford, pp. 293-305.
9. Dando, M. (1999) Benefits and Threats of Developments in Biotechnology and Genetic Engineering, in *SIPRI Yearbook 1999: World Armaments and Disarmament*, Oxford University Press, Oxford, pp. 596-611.
10. Utgoff, V.A. (1997) *Nuclear Weapons and the Deterrence of Biological and Chemical Warfare*, Occasional Paper No.36, The Henry L. Stimson Center, Washington, D.C.
11. Garrigue, H. (2000) Detecting BTW Agents During Military Action, in this volume, pp. 121-128.

12. Eifried, G. (2000) Biotechnological Methods in Detecting Terrorist Attacks, in this volume, pp. 129-140.
13. Pearson, G.S. (2000) Detecting BTW Agents in an Inspection Environment, in this volume, pp.141-159.
14. Tucker, J. (1998) Verification Provisions of the Chemical Weapons Convention and Their Relevance for the Biological Weapons Convention, in *Biological Weapons Proliferation: Reasons for Concern, Courses of Action*, Report No.24, The Henry L. Stimson Center, Washington, D.C., pp.77-105.
15. Garrett, L. (1994) *The Coming Plague: Newly Emerging Diseases in a World Out of Balance*, Farrar, Straus and Giroux, New York.
16. Wilson, H. (1999) BWC Update, *Disarmament Diplomacy* **42**, 27-34.
17. Ad Hoc Group of Governmental Experts (1999) *Report of the 17th Meeting of the AHG*, Document BWC/AD HOC/GROUP/49/Add. 1, 10 Dec.
18. Kelle, A. (1999) Problems and Prospects of CWC Implementation After the Fourth Session of the Conference of States Parties, *Disarmament Diplomacy* **38**, 12-16.
19. Pearson, G.S. (1999) Progress in Geneva. Strengthening the Biological and Toxin Weapons Convention, Quarterly Review No.9, *The CBW Conventions Bulletin* **46**, 5-12.
20. Pearson, G.S. (1996) Improving the Biological Weapons Convention: The Role of Lists and Declarations, in O. Thränert (ed.), *Enhancing the Biological Weapons Convention*, Verlag J.H.W. Dietz Nachfolger, Bonn, pp. 114-139.
21. MacEachin, D. (1998) Routine and Challenge: Two Pillars of Verification, *The CBW Conventions Bulletin* **39**, 1-3.
22. Sims, N. (1988) *The Diplomacy of Biological Disarmament. Vicissitudes of a Treaty in Force, 1975-85*, St. Martin's Press, New York.
23. Pearson, G.S. (1998) The Importance of Distinguishing Between Natural and Other Outbreaks of Disease, paper presented at the NATO ARW on "Scientific and Technical Means of Distinguishing Between Natural and Other Outbreaks of Disease", Prague: Center of Epidemiology and Microbiology, National Institute of Public Health, 18-20 October.

BIOTECHNOLOGY AND THE STRENGTHENING OF THE BTWC

The View from West European Industry

RENÉ VAN SLOTEN
CEFIC
avenue E van Nieuwenhuyse, 4, bte 1
B-1160 Brussels

1. Introduction

The objective of this chapter is to present the views of the FEBC, the Forum for European Bioindustry Coordination, on the question of a Compliance Protocol to the Biological and Toxin Weapons Convention. These views are to a large extent shared by industry in other parts of the world. As I shall explain, FEBC would very much like to become involved in the Geneva negotiation process.

FEBC is an informal grouping for European industry sector federations concerned with biotechnology, which was established to strengthen existing links and to provide a network for the exchange of views and information. FEBC comprises AMEEP (food and food enzymes), CEFIC (chemicals), CIAA (food), COMASSO (plant breeders), EDMA (diagnostic products), ECPA (plant protection products), EFPIA (pharmaceuticals), FAIP (Farm Animal Industrial Platform), FEDESA (animal health products), FEFAC (compound feed), FE-FANA (feed stuff additives), GIBIP (plant and seeds), and EuropaBio (the European Association for Bio-Industries), which provides the Secretariat. FEBC represents directly or indirectly many large, medium or small companies, mainly located in the European Union. Many leading R&D-based pharmaceutical and biotechnical companies in these sectors have only recently started to realize the potential impact of proposed compliance measures and the fact that they may be covered by the scope of the declaration criteria. Industry views are coordinated through the FEBC Task Force Against Biological Weapons, set up in 1996. The group produced a first position paper in August 1997.

Biotechnology is a general term applied to describe the modern use of the combination of the sciences of living things. Its most important element is "genetic engineering", the ability to see and do in the laboratory in a more precise and faster manner what nature has always done slowly and randomly. Biotechnology is not a product, a thing or a production technology but a set of tech-

A. Kelle et al. (eds.), The Role of Biotechnology in Countering BTW Agents, 47–51.
© 2001 Kluwer Academic Publishers. Printed in the Netherlands.

niques used in developing methods for creating products, materials or components.

FEBC members are engaged in the legitimate use of microbiology and newly emerging biotechnologies aimed at the discovery or development of solutions in areas such as health care, agriculture, nutrition and environmental performance. The benefits of biotechnology to society are not always directly visible but include for example:

- New and previously unavailable vaccines, medicines and diagnostics
- Detection of rare and previously undetectable diseases
- New, purer and healthier products
- Lower cost of production giving cheaper products
- Less polluting and less energy consuming production methods.

2.   The Need to Consider the Interests of Industry

It goes without saying that the goals and objectives of the Biological and Toxin Weapons Convention are fully shared by the FEBC. We also support the ongoing endeavors to strengthen compliance with this Convention so as to reduce and hopefully eliminate the threat of biological warfare. The Compliance Protocol is of course first and foremost a matter for the States Parties, with political concerns far outweighing industry matters.

Just as the European chemical industry is not engaged in the production of chemical weapons, the European biotechnology industry has nothing to do with the production of biological weapons. However, we do recognize that many of the R&D and manufacturing technologies and precursors used in our industry could potentially be misused by others to produce such weapons.

FEBC considers it essential that measures strengthening the BTWC ensure that any misuse of microbiology and newly emerging biotechnologies for weapons of mass destruction (or for terrorist purposes) is prevented, *without* impairing legitimate use and continuing development in academia and industry. In this respect I can say that the Chemical Weapons Convention (CWC) managed to strike an acceptable balance between political concerns and industry's interests and as such serves as a model. This does not mean, however, that the declaration and verification provisions of the CWC can be copied directly in the BTWC Compliance Protocol. An important lesson of the CWC has been, however, that it is essential for industry to be able to provide input to the negotiation process. I am reminded of the constructive role the chemical industry was able to play in the development of the CWC.

In our view, controls are acceptable where these are instrumental to the objectives pursued. They must be administratively manageable, not overly burdensome in terms of costs and manpower, safeguard confidential business information and respect the principle of a level playing field. Industry would indeed seriously question the value of any declaration and inspection scheme if this were not applied on a universal basis, resulting in an unfair sharing of the burden and indeed in distortions of competition. It therefore has to be ascertained that the final result of the negotiations will be ratified and implemented by all signatory States Parties.

3. Main Industry Concerns

I would now like to compare two hypothetical scenarios: one in which industry's position is fully accepted by the negotiators and another one which might be called a worst case scenario, in which industry's concerns are completely disregarded. What would industry like to see in the first place?

With regard to a legally binding compliance protocol to the BTWC, FEBC's concerns presently relate to *the scope of the compliance measures* (declarations and inspections) as well as to *the protection of confidential business information* (commercial and proprietary information). Industry's concerns furthermore *include the costs of likely disruption to their activities of an intrusive inspection,* and *the bureaucracy of dealing with complex declaration forms.*

It is clear that the discussions within the Ad Hoc Group of States Parties on a proposed Compliance Protocol have been heavily influenced by the provisions of the Chemical Weapons Convention. However, because of significant differences between research and development into and the production of chemical and biological products, the approach to compliance measures must be a different one. In order to avoid a disproportionately wide coverage of facilities, triggers for declaration under the Protocol must be very precisely analyzed and narrowly defined.

Moreover, primarily because of the ability to remove almost all traces of any development, manufacturing or storage within a short period of time, routine inspections, which are a useful concept under the CWC, are not regarded as a useful concept under the BTWC. The decisive means of compliance control in the BTWC area are considered to be challenge inspections.

As regards declarations, industry recognizes their confidence-building value. The criteria to trigger the declarations presently under consideration, however, include numerous requirements that carry the danger of a disproportionately wide coverage of facilities required to file a declaration. We doubt the effectiveness of such disproportionate declarations in a compliance regime. It is of great importance that individual triggers do not bring into the scope of the regime facilities irrelevant to the purpose and object of the Compliance Protocol, entailing the risk of creating a huge and costly bureaucracy.

With a view to protecting commercially sensitive information, declaration forms should be designed so that there is no possibility of misunderstanding by the company or institution concerned of what is required. Regarding industry's concerns about the risk of loss of confidential business information, it is worth recalling a presentation Dr Imrie of SmithKline Beecham gave at the 1996 NATO Advanced Research Workshop "The Technology of Biological Arms Control and Disarmament" in Budapest, Hungary. This remains a major concern for industry today.

Whilst we accept that challenge inspections are considered a key element of a legally binding Compliance Protocol, I should stress that no member company belonging to FEBC federations is likely to be the legitimate target of such inspections. In our view, short-notice inspections in case of suspected non-compliance should strike the proper balance between the need to try and identify non-compliance with the Convention and the legitimate rights of industry to protect its confidential business information. Therefore, during any inspection strictly managed access must be employed and the inspected site must be entitled to specify what is confidential or proprietary information.

In case of questions and/or ambiguities regarding declarations, clarification procedures are regarded as a useful instrument. Such clarifications might comprise for instance informal consultations and other forms of cooperation with the State Party concerned. Questions and ambiguities should, whenever possible, be resolved by these means. Possible clarification visits in order to clarify and resolve such questions should in any case be the ultima ratio. Any visit to a site to help clear up ambiguities should be under the full control of the site.

4. Optimum versus Worst Case Scenario

If the above cited concerns of industry are duly taken into account and resolved I believe the implementation of the Compliance Protocol will be manageable for industry, will not entail an overly burdensome declaration system, and will meet the concerns regarding protection of confidential business information. Once the States Parties have agreed such a Protocol, industry will wholeheartedly support its implementation, allowing a large number of States Parties not only to sign the Convention but also to ratify and implement it. I would like to refer again to the Chemical Weapons Convention, where industry not only constructively contributed to the realization of the convention but has also subsequently played an active and constructive role in its implementation.

What will happen if industry's concerns are not taken into account? Certain countries could, under pressure from their national industry, decline to ratify the Protocol / Convention, thus preventing the creation of a level playing field. Obviously this would not only undermine the success of the Protocol altogether, since universality is a precondition of its success, it would also mean industries in countries that have ratified the Protocol / Convention would face the burden of such a regime while industries in other competitor countries would not. This might lead to changing investment patterns or even relocation of industries.

In conclusion, a proper dialogue between the States Parties and industry is needed to ensure that the right balance is struck between the interests of both sides.

FEBC is prepared to offer its expertise and experience across all the sectors it covers to help design an instrument that will provide the depth and scope necessary to deter the misuse of microbiology and biotechnologies while at the same time preserving the legitimate rights of industry.

STRENGTHENING THE BTWC THROUGH R&D RESTRUCTURING: THE CASE OF THE STATE RESEARCH CENTER OF VIROLOGY AND BIOTECHNOLOGY "VECTOR"

LEV S. SANDAKHCHIEV AND SERGEY V. NETESOV
State Research Center of Virology and
Biotechnology VECTOR
Koltsovo, Novosibirsk Region
Russia

1. General Characteristics of the State Research Center of Virology and Biotechnology "Vector"

In the Siberian region of Russia, there are several facilities and factories that are considered to be related to the Initiatives for Proliferation Prevention Program. The State Research Center of Virology and Biotechnology "Vector" is one of them.

Though the Center has not been directly involved in BW design or production, before 1989 it was primarily oriented towards defensive programs research and up to 90% of its research funding came from these programs.

During the period 1974-1991, the Center created an advanced scientific base allowing research on highly dangerous infections at Biosafety Level-3 (BL-3) and BL-4, theoretical and experimental studies in the area of genetic engineering, biotechnology and aerobiology.

The staff of our facility totals some 1,200 researchers, including 4 academicians, 17 Doctors of Science and 166 PhDs.

The total area of our research and production facilities amounts to more than 250 thousand square meters; our property covers 8,000 hectares (19,768 acres). Our main research and production area located at Koltsovo has some 14,000 residents. Our facilities have all the necessary infrastructure, including transportation, electricity, water and energy, sewage treatment, etc. We also have sufficient vacant land to support the construction of new facilities.

Prior to 1991, our Center was purely a science and technology complex with almost no production, and it received all of its funding from the federal budget.

A. Kelle et al. (eds.), The Role of Biotechnology in Countering BTW Agents, 53–59.
© *2001 Kluwer Academic Publishers. Printed in the Netherlands.*

2. "Vector" R&D Restructuring

After the dramatic decrease in defense-related projects in 1991 and general cuts in state funding due to the economic situation in Russia, the Center had to reconsider and redirect its R&D activity towards basic and applied problems of public health, veterinary medicine, and environmental protection, as well as to begin production of medical and diagnostic products for public health and veterinary medicine.

The main goals of VECTOR R&D restructuring are:

- Reorient our personnel with extensive experience in the areas of virology, genetic engineering, and biotechnology to carry out research for the needs of public health, agriculture and environmental protection.
- Carry out basic research in the area of molecular virology and virus mutability.
- Enhance applied research aimed at the development of new therapeutic, preventive, and diagnostic agents for public health and veterinary medicine, employing the latest advances in genetic engineering and biotechnology.
- Construct production facilities to manufacture a wide range of drugs for public health and veterinary medicine.
- Collaborate internationally both in fundamental and applied research and in drug production.

To achieve the goals we have carried out a structural reorganization of the Center by assigning research institutes to specific areas of basic and applied studies, establishing production and marketing facilities, and adapting service and support facilities to these structures.

3. "Vector" Production Facilities

During the last few years the Center has managed to create its own production facilities for producing different diagnostic kits, vaccines, immunobiological preparations, laboratory-use products, etc. numbering more than 150 items. Most of these preparations were designed at our Center.

The reconstruction of our facilities has enabled "Vector" to establish several advanced production sites for the cultivation, purification, and manufacture of dry and liquid injectable drugs, as well as pre-packaging sites (blister and vial packaging) for ready-to-use drugs.

In addition, we have been efficiently cooperating with some well-known pharmaceutical companies from abroad such as APOTEX (Canada), Bayer AG (Germany), Norton Healthcare (Great Britain), KRKA (Slovenia), LEK (Slovenia), POLFA (Poland), manufacturing drugs from substances imported to Russia. To be exact, currently this is restricted to the final stages of drug production: pre-packing and quality control. Even so, this has become a very profitable business because this makes it possible to identify and promote the most promising drugs to the Russian market at a fairly low cost.

In order to improve further the production facilities we shall have to undertake the following:

- reach the level of international GMP certification;
- expand international cooperation with pharmaceutical companies on early stages of production of substances;
- facilitate innovative projects to produce new vaccines and treatments that are in demand on the domestic and international market.

Our Center is one of the ten most efficient and dynamically developing manufacturers in Russia in terms of sales volume and nomenclature of drugs. We operate throughout Russia via our efficient network of distributors. Thus our own production and sales are making "Vector" less dependent on state funding.

4. Funding Sources and Income Structure

The proportion of our income obtained from product sales has grown significantly because of production growth. Federal funding has been radically redirected from defense-related projects towards fundamental and applied research for human applications. The largest part of our funding comes from the Russian Ministry of Science and Technologies. Funding from other sources is relatively small. However, we highly appreciate any international efforts aimed at supporting science in Russia, so I would like to stress that we collaborate productively with the International Science and Technology Center (ISTC), and we have grants from CRDF, INTAS and some others as well. Product sales account for 72 % of our total income, of which 70 % comes from sales of drugs that are jointly produced with foreign companies.

5. International Cooperation

In the area of virology and molecular biology, we actively collaborate with 12 leading research centers abroad, including the Centers for Disease Control and Prevention in Atlanta, USA; the National Institute for Allergy and Infectious Diseases, NIH, Bethesda, USA; Institute of Virology, Marburg; and Institute of Tropical Diseases, Hamburg. Our Program for Research and Monitoring of Pathogens of Global Importance was initiated in 1998. Since 1994, our Center has housed the Russian Collection of Variola Virus Strains. Recently our facility was officially designated as a WHO Collaborating Center for Orthopoxvirus Diagnosis and Repository for Variola Virus Strains and DNA.

More than forty of our scientists are currently working at several research institutions abroad, and this gives us an excellent opportunity to establish closer contacts with these facilities. Both at home and abroad, our scientists are now working in the most advanced area of biotechnology: phage display. Regarding our applied research, we have some almost complete R&D projects related to diagnostics, vaccines, and cytokines production that may be interesting to our potential partners.

"Vector" is also interested in participating in further international programs such as WHO programs on developing new vaccines, INTAS, Copernicus, EUREKA, NATO Scientific Committee programs, ProCEID, and we are prepared to discuss with any potential partners other projects that may be considered as proposals for bilateral or international cooperation or projects ready for direct commercial application.

6. Issues of Transparency and BWC Strengthening

First of all, there should be no doubt about the determination of the new Russia to promote international cooperation, especially cooperation aimed at the prevention of proliferation involving biological and toxin weapons.

In April 1992, the Russian President signed Decree #390 prohibiting on the territory of Russia the conduct of any work that would violate the Biological Weapons Convention. The appropriate articles were included in the Russian Criminal Code. Import/export operations with pathogenic microorganisms are now subject to licensing. A law has also been approved that provides for government regulation of any genetic engineering research.

Our Center annually reports to the UN on the use of BL-3 and BL-4 laboratories involved in research on highly dangerous pathogens. The work in this area is also subject to control by the Russian Committee on Issues Relating to the BW Convention in the Office of the Russian President. All this demonstrates that the political and legal environment in Russia is completely consistent with the proliferation prevention concept. Beyond this, however, our main efforts must be directed towards the involvement of specialists and production capabilities in civil applications, and the prevention of the outflow of researchers and technologies to places with an unstable political environment.

Our Center has managed, with governmental support, to re-specialize its scientific activities and initiate, though not on a large scale, its own production. However, these efforts are insufficient and continue to leave our scientists badly underpaid in comparison to their colleagues in developed countries.

We believe that the most satisfactory concept of international collaboration in the area of dangerous pathogens research was developed in 1997 by an expert group of the US National Academy of Sciences (NAS) under the leadership of Prof. Joshua Lederberg and Dr. John Steinbruner [1]. This document was prepared with the involvement of Russian experts, and it seems to be the most serious analysis of all pros and cons arising from collaboration in dangerous pathogens research.

In the NAS concept it is envisaged that confidence building measures and transparency be provided in a broader context. In particular, it is envisaged that in each specific project regular access should be provided to the place of performance and the necessary information be provided in compliance with the principle of extending active guarantees to counterparts. It is also envisaged that collaborative work between Russian and American scientists be carried out in laboratories in these countries under a separate agreement. Freedom of access is more vaguely defined in such international programs or foundations as ISTC, CRDF, NASA, DARPA and IPP (DOE), which we can quite understand since the declared goal of these programs is to support activities not involving dual use technologies.

The experience of our Center shows that efforts funded by the abovementioned programs involve only part of our personnel and do not fully employ the most qualified staff and the high containment laboratories which, however, are of greatest concern to experts. 6 pilot projects concluded under the NAS initiative have involved only 5 % of the staff and high containment available.

Recently, we have been actively discussing this issue with experts from the Russian Ministry of Public Health the Russian Ministry of Science and Technologies, as well as with our colleagues from different US agencies, ISTC, and the International Research Center "The Joint Institute for Nuclear Research" in Dubna, Moscow Oblast.

We believe it serves the interests of non-proliferation and world-wide public health to create in Russia *International Centers* for the study of pathogens of viral (Koltsovo) and bacterial (Obolensk) nature as a model for work on dangerous pathogens. We believe it is very important that WHO as an international public health authority be involved early in this process, to participate in discussing the feasibility and developing the concept of the proposed *International Centers*.

Under the term *International Center* we understand an international organization established by an intergovernmental agreement like those of the already existing ISTC or the Joint Institute for Nuclear Research in Dubna. Only this context will make it possible to achieve all goals of non-proliferation and threat reduction by ensuring the fulfillment of major requirements for transparency and confidence building: free access to the program and the results obtained, and free access to all facilities and all staff of the *International Center*. A continuous involvement of foreign scientists in work on site is a powerful instrument of confidence building. Therefore, we believe it is critical to include all high containment capabilities and supporting facilities in the *International Center* in order to alleviate concerns over possible illicit activity. It may make sense to use documents provided by the Dubna Institute in drawing up charter documents for the proposed *International Centers*.

The Dubna Joint Institute for Nuclear Research employs 6,000 staff. Of these 4,000 receive support from the Institute's budget. Of these 1,000 are research staff. Of these 1,000 research staff, 400 are foreign researchers working on a contract basis (3 to 5 years).

Another international center that uses a similar model is CERN in Switzerland. The reputation enjoyed by these institutes, both of which work in the highly sensitive area of nuclear physics research, is indicated by the fact that both of them are under consideration for the Nobel Peace Prize.

Our Center also has ideas concerning the main objects, goals and directions of the research of the proposed *International Center*.

We would like to suggest concentrating research on arboviruses, including tick-borne encephalitis virus which is endemic in Russia; HFRS virus; Omsk hemorrhagic fever virus; filoviruses: Marburg and Ebola; orthopoxviruses: smallpox virus, monkeypox, and cowpox viruses; viruses causing hepatitis A, B, and C; paramyxoviruses, rabdoviruses, and influenza viruses. This list of viruses could be extended by tuberculosis and opistorchiasis – a human parasitic disease which affects the liver. The latter is endemic in Siberia. It also makes sense to address viruses causing zoonotic and anthropozoonotic diseases critical to public and veterinary health.

The scientific programs of the proposed *International Center* could be based on recommendations contained in the Controlling Dangerous Pathogens report of the US NAS [1]. Research areas should cover fundamental aspects of genetics, physiology, and biochemistry; pathogenesis studies, including those on human immune response to infectious diseases; development of diagnostic methods; design of drugs and vaccines; epidemiological studies, including the investigation of environmental factors and their effect on rodents and insects; the effect of human behavior and demography.

To my mind, it is also important to carefully address the issue of enrolling the staff in various training and exchange programs. It is critical to also consider the issue of appropriate handling of microorganisms.

Special attention would be paid to the investigation of outbreaks of infectious diseases in the region (the Asian part of Russia, Mongolia, the Central Asian members of the C.I.S.; possibly China). This investigation could be conducted using the molecular epidemiology approach, which allows us to determine the sero- and genotypes of infectious agents and the possible source of primary infection. Such investigations could be carried out on a regular basis for a wide range of pathogens, which would be extremely useful for monitoring the evolution of infectious agents. The proposed *International Centers* will be able to provide modern fast diagnostics and monitoring of infectious disease agents in the vast territory of East Europe and Northern Asia, which would be extremely helpful for prognosis of the evolution of infections endemic to this region and of whatever new infections might emerge in this region in the near future.

In addition, the creation of the proposed *International Centers* will make it possible to unite our efforts to gain knowledge that is essential to control potential bioterrorists. It is, however, important to establish a regime for the appropriate dissemination and use of the scientific results obtained that might facilitate the possible misuse of biological agents.

Of course, the proposal we have presented here is only a brief description of this idea, but we hope to develop it in a detailed document in the near future.

We strongly believe that with the political and financial support of all the NATO governments and Russia we will be able to fight against dangerous pathogens together, in full confidence that our collaboration will be efficient and fruitful for the benefit of all our nations.

7. References

1. National Academy of Sciences (ed.) (1997) *Controlling Dangerous Pathogens: A Blueprint for US-Russian Cooperation*, National Academy Press, Washington, D.C.

DISSOLUTION OF THE SOVIET UNION, INTRODUCTION OF A MARKET ECONOMY AND THE FUTURE BTWC COMPLIANCE PROTOCOL: IMPACT ON THE RUSSIAN BIOTECHNOLOGY INDUSTRY

JOHN COMPTON [1]
International Science and Technology Center
Luganskaya ulitsa, 9
Moscow
Russian Federation

1. Introduction

As a consequence of the dissolution of the Soviet Union and end of the Cold War in the early 1990s Russian science has suffered severe setbacks in terms of government funding and support. This has led to a severe loss of personnel, reduction in the ability to do research, and even the closing of some institutes. Most science was connected in some way or another to defense spending, and the vast network of scientific institutes employed the country's best minds towards primarily military ends. The biotechnology field has suffered along with other defense-related areas such as nuclear science, chemistry and materials science, and rocketry. The effect on the future of the biotechnology industry has been indirect but nevertheless profound, because it is only the scientific institutes with their concentration of equipment and talent that can provide the driving force behind new developments and applications in the manufacturing sector. This is, of course, in stark contrast to countries which have well-developed private industrial capacity and extensive private research laboratories. The more immediate factor that has crippled the Russian biotechnology industry is the wholesale switch of the consumer market from domestic supplies to imported medical products in spite of the much higher prices for the imports. For example, whereas Russia once produced its own insulin, it now relies on imports alone, a fact which has put many diabetics in peril after the dramatic fall of the ruble with the financial crisis of August 1998. More telling is the fact that in a typical pharmacy in Moscow, the majority of products are imported, as demanded by a public that frequently perceives Russian products, rightly or wrongly, as inferior to their western counterparts.

61

A. Kelle et al. (eds.), The Role of Biotechnology in Countering BTW Agents, 61–67.
© 2001 *Kluwer Academic Publishers. Printed in the Netherlands.*

2.   Effects of Perestroika

In order to put the effects of perestroika on Russian industry into perspective it is necessary to compare the Soviet system with today's. The switch from a centrally-planned economy to a form of capitalism with the still noticeably heavy, if unsteady, hand of government has been most profound. Unbridled capitalism (as, for example, in Hong Kong), shocking as it might have been to Russian industry, would be greatly preferable to today's system where corruption and ineptitude damage both the well-being of the economy and the interests of the consumer. The unstable situation with regard to banks, currency, the government's payment of its obligations, and tax collection leaves Russia with a financial environment that is not conducive to private investment or consumer confidence. But this is only one aspect of the troubles faced by the domestic biotechnology industry in Russia.

Corruption in the licensing and procurement procedures for medical products such as pharmaceuticals and vaccines has totally favored the well-financed foreign multinational corporations over the biology institutes that have difficulty in making payroll and utility payments. In Soviet times scientists held secure lifetime positions with privileges and prestige enjoyed by few citizens other than the government elite. Now they are often forced to choose between accepting a minimal salary with poor working conditions, emigration, or taking on a second job, if they wish to carry on with work in their chosen field.

3.   Value of Russian Expertise in Biology

The Soviet Union had an enviable record of setting up a network of hundreds of superb biological institutes spanning a host of specialized lines of investigation. These represent a true national treasure in terms of their human talent and capacity for carrying out world-class research. In spite of losses of personnel due to attrition and emigration, they still have sufficient staff and equipment to carry on as in previous times, provided that funding is available. Most of those who have left Russia were either well-known in the international scientific community or were young, energetic, and had a knowledge of English so that they could adapt readily to a new culture and working environment. Because the Russian biology institutes have relatively low salaries and operating costs, they represent a real bargain for foreign companies that want to locate part of their research abroad. The savings can be staggering when one considers that a Russian biologist can be supported for some fifty times less than his western counterpart. In spite of this advantage, the desire for secrecy and fear of instability on the part of the highly competitive foreign corporations limits their desire to invest in Russia. Similarly, production costs for manufactured goods are considerably lower than abroad. However, typically the foreign firms are willing to allow only the repackaging and marketing of their products on Russian soil.

4. Slow Progress in Commercialization

Many early hopes for the commercialization of native Russian biotechnology have remained unfulfilled. The biotechnology institutes planned to spin off a series of commercial ventures with the help of foreign investment capital, but in most cases have had to settle for more modest projects of limited scope and duration. This underlines a traditional weak point in Russian science in general, namely, the large gulf between pure science and practical application. This is a relic of the Soviet system, which despite concerted efforts to find applications for science failed to provide sufficient mechanisms and incentives to bridge the gap between research and production effectively. In addition, Russian scientists rarely had contacts with their colleagues in other parts of the world. Only government officials and intelligence experts were allowed to develop contacts abroad. Likewise, within the Soviet Union official secrecy isolated scientists not only from industry but also from each other.

Several additional factors combine to make the road to commercialization of Russian biotechnology a difficult one. The practice of keeping accurate laboratory records in order to safeguard intellectual property rights (IPR) is a difficult one to instill in researchers who were never before burdened with such details. Other IPR issues tend to be perceived as alien and need to be taught intensively, as in a classroom setting. This lack of familiarity with international business practices has in some cases led to the exploitation of Russian scientists by unscrupulous western businessmen. News of such unfortunate events spreads quickly throughout the scientific community in Russia and tends to discourage further contacts with the West. Russian government corruption has already been mentioned as a serious obstacle to developing a free-market economy. Confiscatory tax rates, unevenly and haphazardly enforced by a welter of different authorities, and even retroactively applied taxes, also tend to destroy incentives by denying profits to those who would be entrepreneurs. Fortunately, by clever means these problems can be avoided in part, and indeed tax evasion has become a necessary survival skill for many Russians. However, the larger the enterprise, the more difficult it is to escape the attention of the tax authorities.

5. Research Favored over Production

Several factors conspire to limit foreign investment in large-scale production in biotechnology in Russia. Risk is largely responsible for cultivating a climate of reluctance on the part of potential investors. There are the more obvious financial risks associated with a shaky economy and an uncertain political future. But there are deeper currents at play as well. Russian civil law is very poorly developed and will require years to mature. The tax system has also been mentioned as a major problem. Tied to all this is a fear of loss of intellectual property rights. This is partly due to the lack of recourse to civil law in order to protect these rights, but also partly to the lack of a practical tradition that protects intellectual property.

Smaller research contracts with Russian institutes avoid many of the risks mentioned above and so are more attractive to potential investors in Russian biotechnology. The huge savings in labor costs are the driving force behind such agreements and help offset the barriers to cooperation that are presented by the prevailing harsh Russian business climate. Larger investments in production-scale enterprises are not so attractive because facilities meeting international standards of health, safety, and sanitary conditions are rare in Russia. Lastly, it should be mentioned that the lack of history of a market-driven economy in Russia has not created a climate conducive to contacts with the more industrialized countries and has not instilled the entrepreneurial spirit in the native research community.

Foreign-sponsored research is not without its problems, however. Much of the business knowledge and skills taken for granted in potential investor countries is conspicuously absent in Russia. There is simply no history of keeping accurate records of intellectual discoveries made in the course of research, for example by signing and witnessing laboratory notebooks on a regular basis as is commonly done in countries with strong private industry. There is also a tendency on the part of investing enterprises that even a leak of information concerning the line of investigation being pursued could be critically damaging if it fell into the hands of competitors. Even where economic concerns are minimized, the question of secrecy weighs heavily on the minds of investors. They also tend to perceive the Russian scientists as out of touch with the latest areas of profitable research, a situation fostered by the relative lack of market-driven investigation. Also the Russians often have difficulty in advertising their own talents to foreigners. Superficial as well-scripted presentations and fancy promotional literature may seem, they tend to engender a sense of confidence in prospective business partners. Education of research staff in international business practices, notably intellectual property rights, is essential for Russian scientists.

6. Effects of BTWC Compliance

It has already been mentioned that Russian physical facilities are seldom up to international standards for production in the biotechnology field. It is often the case that entirely new buildings are cheaper to construct than older facilities are to refurbish. This is because international standards for biotechnology are quite tough and also require relatively advanced technology and equipment. This extends even to the handling of experimental animals, and does not entail simply their care and housing. Extensive and systematic documentation of animals is necessary for adherence to administrative rules and for prevention of any political scandals that might arise out of so touchy a subject as humanitarian treatment of those animals. Inspections, a rather new phenomenon in Russia biotech facilities, are another area that require proper bookkeeping and a grounding in rules and procedures. A thorough education for research institute staffs on compliance with international research and production standards is an important step in meeting the challenges imposed by BTWC compliance.

7. Effects of the Economic Crisis of August 1998

The financial crisis of August 1998 turned an already risky Russian business climate into a rather ominous one fraught with many bank failures and other dislocations. Thousands of foreigners decided to close their operations forever and left the country in a rush so sudden that long waiting lists grew at the international moving firms. The effect on foreign investment was devastating as millions of dollars invested in Russia were lost forever. The effects within the country were no less striking. Many people lost their entire savings (if they had not already done so on one of several previous occasions in the past decade when a devaluation or manipulation of the currency occurred). In other cases they were unable to withdraw their money from their accounts, or even receive their salaries which had been freshly deposited in some particular bank. Many who had never fully considered emigration in the past were pushed over the brink and seriously began to look for new lives abroad. Scientific institutes faced frozen accounts and were unable to pay bills or salaries. Fortunately, some of the effects of the crisis eased by early 1999, and monetary transactions began to return to normal at those banks that had survived the crisis. But the psychological effects are not so easy to erase, and such upheavals serve to make already cautious investors evaluate the risks of supporting Russian scientists even more carefully than before. Some signs of political change and various rumors only exacerbate the situation. For example, reports circulated by word of mouth that the government intended to restrict foreign access to scientific institutes, while at the same time not increasing the trickle of money flowing to them, thus dooming some to collapse. Fortunately, government inaction in this case has been a blessing that falls into the category of benign neglect. However, military actions by NATO, for example, could prove the catalyst to limit foreign cooperation in non-proliferation of expertise in weapons of mass destruction.

8. Survival of the Biology Institutes

The Russian biology institutes have been developing various strategies for maintaining operations as research centers, which entails holding onto remaining staff, keeping at least some buildings physically operating, and permitting research to continue. Some methods of earning money from Russian sources (such as manufacturing bottle caps, providing laundry services for others, and incinerating waste paper) in facilities designed for support of biological research are particularly illustrative of the gravity of the situation. Unfortunately, in those cases where commercial spin-offs have been successful, there is not necessarily a financial benefit to the parent institution. Private support for research projects from abroad also accounts for some income at the institutes. More often private support amounts only to repackaging of imported products for the Russian market.

Considerable interest from abroad in non-proliferation of Russian biotechnology has generated enough capital from governmental and non-governmental agencies to support some projects in some institutes. However, the total amount spent is far lower than what the Soviet government could afford in pre-perestroika times and can only be regarded as a palliative and piecemeal solution to an overwhelming need. Fortunately, this foreign support has been growing in recent years and amounts to millions of US dollars. For example, the Defense Advanced Research Projects Agency (DARPA) of the US Department of Defense has recently committed itself to supporting a number of Russian biological institutes with substantial and quite forward-looking projects.

9. Future Needs for the Growth of the Russian Biotechnology Industry

The requirements for the survival of the Russian biotechnology industry are quite unambiguous. Further integration of Russian biologists into the world community is essential as a fundamental prerequisite. This can be accomplished by building contacts at international conferences and through collaboration in cooperative research projects. These contacts then help to build a stronger orientation in Russian biologists towards applied research and market-driven commercialization of results. Also, they promote adoption of international standards for manufacturing, business practices, and management of intellectual property rights. The International Science and Technology Center in Moscow is encouraging all such activities and facilitates foreign aid through tax advantages and direct payment of Russian scientists, all of which mitigates the risk factors of the business environment for contract research activities.

It is clear that substantial foreign investment will be required for revitalization of the anemic Russian biotechnology industry. The Russian government can cooperate by adopting a fair and rational scheme of taxation, strengthening the civil justice system, and promoting a healthy financial climate within the country. So far the indications are that implementation of the above requirements will be a slow and difficult process. But at least this is all underway, and the pace of improvement is accelerating modestly in the face of formidable barriers to progress.

10. References

1. The material for this report was gathered from interactions with a series of Russian biological institutes, including interviews with their directors and other staff members. Most information came from more than one independent source or is an amalgam of impressions gained in the course of day-to-day work and in informal conversations with colleagues and clients.

THE POTENTIAL OF DIFFERENT BIOTECHNOLOGY METHODS IN BTW AGENT DETECTION

Antibody based methods

PETER SVESHNIKOV
Russian Research Center for
Molecular Diagnostics and Therapy
Sympheropolsky Boulevard 8
13149 Moscow
Russia

1.   Introduction

The highly sensitive detection of CW and BTW agents, namely chemical compounds, biomolecules (toxins) and microbiological agents (bacteria, viruses, etc.) with the use of antibodies is based on the transformation of a specific biomolecular interaction between antigen and antibody into a macroscopically detectable signal or a change in the physical properties of the media. Immunochemical methods have been successfully applied in the detection of pathogens, or the antibodies produced in response to these pathogens, for more than 50 years. During this time a great number of laboratory techniques have been developed [1].

Generally speaking, there are two ways to increase antibody-based assay sensitivity and specificity. The first is to obtain high affinity antibody and develop molecular systems to enhance the specific interaction of antigen and antibody. The most impressive examples in this respect are the development of hybridoma technologies for the production of monoclonal antibodies with predetermined properties, as well as the application of multienzyme cyclic systems, avidin-biotin complexes, enhanced chemiluminescence and time-resolved fluorimetric technique to aid in the assay of antigen-antibody interactions. The second is to use systems of signal registration that are capable of detecting single molecular interactions. In this case practically absolute sensitivity is achieved, and single molecules and particles define a detection limit.

The purpose of this chapter is to demonstrate the potential of antibody-based biotechnology methods for the detection of BTW agents.

69

A. Kelle et al. (eds.), The Role of Biotechnology in Countering BTW Agents, 69–77.
© *2001 Kluwer Academic Publishers. Printed in the Netherlands.*

2.   Immunoassay parameters

The list of parameters which form the basis of the immunoassay optimization approach is presented in Table 1.

The main parameters are the sensitivity, specificity and duration of the analysis. As a rule, the sensitivity of immunoassays is very high, achieving detection of substances in the concentration range of 5-10 pg/ml. Depending on antibody specificity, one can detect low molecular weight substances, biomolecules (proteins, components of microbial cell walls), viruses and whole bacteria, and the identification may be generic or strain-specific. By this parameter immunoassays are somewhat less sensitive than procedures leading to the isolation of the pathogen in culture, but the specificity is nevertheless acceptable and the procedure is usually easier to perform and can be carried out more rapidly.

TABLE 1. Immunoassay performance characteristics

Parameter	Range
1. Sensitivity	10-1000 pg/ml $10^3 - 10^5$ particles/ml
2. Specificity	85 – 100 % strain-generic
3. Time	5 – 120 min
4. Sample volume	5 – 10000 µl
5. Biomaterial	Blood, plasma, serum, urine, saliva, stool, tissue extracts, water, soil, air etc.
6. Stability	6 – 12 months
7. Versatility	Single species-multiparameter
8. Automation	Manual, hand-held robotized, fully automated work stations

The time of assay is determined by its design (format) and varies in the range from 5–10 min to 2 hours. Usually sensitivity must be sacrificed for the sake of time reduction. The main requirements are that the sample should be in aqueous solution, usually a physiological saline buffer having a neutral pH (7±1). Also, the reaction temperature should be about 25±10°C. Strong acids and alkali, organic solvents, solid and gaseous substances cannot be subjected to immunoassay without preliminary treatment. Protein molecules (antibodies and their derivatives) are a major component of test-kits. Therefore, the shelf life of the kit is determined by the stability of the antibodies and usually ranges from 6–12 months when the kit is stored at +4°C. Immunoassay results may be qualitative or quantitative. Qualitative results may often be detected with the naked eye, but quantitation normally requires measurement using instruments such as a photometer, fluorimeter, luminometer or similar equipment. In the majority of cases the kits are developed to determine one parameter, but recently, new versions of multi-parameter detection kits have become available. Technically, the design of enzyme immunoassays may vary greatly: they may be packaged as miniature disposable devices for use in the field or large automated machines capable of performing thousands of tests per day based on different assay parameters in laboratories. The above characteristics are general and valid for practically all immunoassays. It should be noted that the most widely used techniques are: enzyme-linked immunosorbent assay (ELISA), time-resolved fluoroimmunoassay (TR-FIA), latex-agglutination, immunochromatography, immunofiltration, flow cytometry, and magnetic bead separation based immunoassays. Objectively speaking, it is difficult to express a preference for the one or the other method, because each has its own merits and drawbacks. Some of these methods will be dealt with later on in this contribution.

2.1 THE QUALITY OF THE ANTIBODIES REQUIRED FOR THE IMMUNOASSAY

The main component of all immunoassays, which provides for the specific biological recognition of the analyte, is the antibody. Antibodies are distinguished by their source as well as the method used to obtain them and can be poly- or monoclonal. Polyclonal antibodies are obtained from the serum of hyperimmune animals. Monoclonal antibodies (MAb) are produced by hybridomas, which are products of the fusion of myeloma cancer cells and single splenocytes (antibody-producing cells) [2]. Modern technologies have made the large scale production of MAb possible, either in bioreactors or in mice with ascitic tumors. As a rule, polyclonal antibodies demonstrate a higher affinity for antigen and wider specificity as compared to MAb. MAb are more standard preparations and very often make it possible to detect extremely small differences in the structure of infectious agents. In general, the requirements that antibodies should meet depend on the purpose of the kit and its design: antibodies should be of high affinity and precisely determined specificity, and their activity should be preserved upon immobilization on a solid phase and after conjugation with enzymes or other labels. For an optimal selection of antibodies, thorough and differential, extended studies are required.

Our Research Center for Molecular Diagnostics and Therapy (RCMDT) has extensive experience in the production of MAb to infectious agents, including those of special interest from the point of view of the BTWC. Properties of MAb are presented in Table 2.

TABLE 2. Monoclonal antibodies infectious disease panel

Name	Prod.Year	Specificity
1. Influenza Virus Type A	1981	HA, NP
2. Herpes Simplex Virus Type II	1983	gD
3. Rotavirus	1984	p42inner capsid ag
4. Influenza Virus Type B	1985	type-specific ag
5. Adenovirus	1986	hexon ag
6. Rabies Virus	1987	gP, NC
7. Respiratory Syncytial Virus	1988	F-protein
8. Hepatitis B Virus	1989	HbsAg, "a"epitope
9. Epstein-Barr Virus	1990	CA: pl20 gP160
10. Hepatitis B Virus	1995	HbcAg, match pair
11. Rubella Virus	1997	E1, E2, C (VLP)
12. Mumps Virus	1998	Under development
13. Hepatitis A Virus	1998	Panel
14. Yersinia pestis	1982	F1 antigen
15. Bacillus anthracis	1985	Protective Ag
16. Francisella tularensis	1987	LPS
17. Salmonella spp.	1990	A,B,D group O-Ag
18. Legionella pneumophila	1992	LPS
19. Toxoplasma gondii	1997	P30
20. Chlamydia trachomatis	1997	P67
21. Candida albicans	1997	LPS
22. Burkholderia mallei	1998	Under development
23. Cholera toxin	1983	α-, β-subunits
24. S.aureus enterotxin B	1987	Panel
25. Diphtheria toxin	1989	Panel
26. Clostridium botulinum D-toxoid	1998	Panel
27. Clostridium botulinum A,C-toxoids	1998	Under development

Among viral pathogens, a number of MAb to agents of respiratory diseases, i.e. influenza viruses A and B, Rous sarcoma virus (RSV) and adenovirus, have been produced. These antibodies have high diagnostic potential in ELISA and indirect immunofluorescence. In diagnostics of gastroenteritis and diarrhea we use MAb against rotavirus and adenovirus, which allow detection of viruses in feces. MAb against hepatitis A and B viruses are used in enzyme immunoassays as well as in sample enrichment for the polymerase chain reaction (PCR). MAb to the glycoprotein and the nucleocapsid protein of rabies virus are of special interest, because they allow precise typing of different isolates as well as the neutralization of rabies virus in vivo. A panel of anti-microbial MAb is directed against several pathogens of extremely dangerous diseases such as plague, tularemia, and anthrax. These antibodies are used in ELISA, agglutination tests and immunofluorescence assays. MAb to *Salmonella* group-specific O-antigens and the main plasma membrane protein of *Toxoplasma gondii* are presented, with numerous examples for application in clinical practice. Among anti-toxin MAb a panel directed against *Staphylococcus aureus* enterotoxin B should be mentioned [3], as well as new antibodies against the botulinum toxin family, which can find application in food safety programs.

In my view the problem of supplying immunodiagnostic kits with high quality, standardized and specific antibodies can only be solved in the framework of international co-operation, as the variety of known and emerging infectious agents is enormous and exceeds the analytical capacities of individual laboratories.

3. Immunofiltration

The state of the art in the modern biomedical sciences is defining new goals in the field of immunodiagnostics, which are aimed at increasing sensitivity and reducing assay time. For the purposes of achieving these goals, MAb, non-equilibrium conditions for antigen-antibody complex formation (especially flow-through techniques), and polymeric indicator dyes have found wide application.

In recent years at RCMDT a new technology of express-diagnostics for infectious agents has been developed using the methods summarized above. To shorten the immunoassay time an immunofiltration analysis principle has been applied, based on large pore type polymer filters or non-porous glass microspheres, which, when packed in microcolumns, provide a liquid flow rate no less than 0.1 ml/min. Sensitivity and specificity of the assay are maintained on two levels: first, by using a large excess of MAb for immobilization on a solid phase to serve as an immunoadsorbant for antigen capturing, and second, by using biotin-labeled MAb as secondary antibodies, which are introduced to the assay after the tested sample has been applied to the column. The high sensitivity of the method is guaranteed by three preconditions: (i) the large capacity of immunosorbent; (ii) the possibility of accumulating an excess of secondary biotinilated MAb; (iii) use of the polymer conjugates streptavidin-albumin-dye (STALD) for visualization of immune complexes with subsequent photometric detection. The scheme of the immunofiltration analysis is presented in Fig.1.

In addition to photometric detection, fluorescein-labeled secondary antibodies can be used to measure antigen-antibody reactions. This removes one stage from the assay time and increases the sensitivity 10-fold. At the same time, however, this type of measurement requires a more sophisticated instrumentation designed to detect immunofluorescence. We plan to use FITC-labeled antigen fragments for continuous monitoring of the presence of pathogens. Potentially, the assay time can be further reduced by using portable photometers and fluorimeters, which allow measurement of the signal directly on

74

the column without elution of the dye. The characteristics presented in Table 3 sum up the merits of the suggested technology.

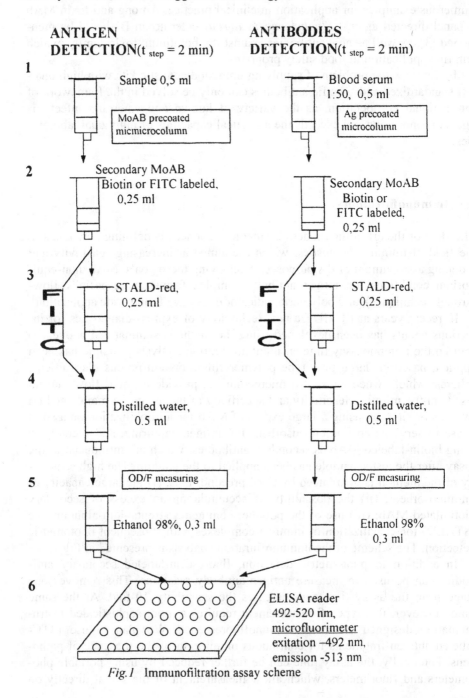

ANTIGEN DETECTION(t_{step} = 2 min)   **ANTIBODIES DETECTION**(t_{step} = 2 min)

1   Sample 0,5 ml   Blood serum 1:50, 0,5 ml

MoAB precoated micmicrocolumn   Ag precoated microcolumn

2   Secondary MoAB Biotin or FITC labeled, 0,25 ml   Secondary MoAB Biotin or FITC labeled, 0,25 ml

3   F I T C   STALD-red, 0,25 ml   F I T C   STALD-red, 0,25 ml

4   Distilled water, 0.5 ml   Distilled water, 0.5 ml

5   OD/F measuring   OD/F measuring

Ethanol 98%, 0,3 ml   Ethanol 98% 0,3 ml

6   ELISA reader 492-520 nm, <u>microfluorimeter</u> exitation –492 nm, emission -513 nm

Fig.1 Immunofiltration assay scheme

TABLE 3. Main parameters of immunofiltration assay

Time of assay from the moment of the sample administration to the obtaining of the result	10 – 20 min
Number of stages including sample preparation	4 – 6
Assessment of the results	visual / instrumental
Instrumental detection	photometry / fluorimetry
Type of assay	quantitative / qualitative
Sensitivity	0.1 – 10 ng/ml
Sample volume	0.25 – 10 ml
Specificity	not less than 98 %
Variation coefficient	8 – 15 %
Number of tests	1 – ∞
Type of sample	whole blood, serum, plasma, urine, feces, nasal washings, cell culture supernatants, water-salt and organic extracts and washings, drinking water, industrial drainage waste water
Stability	min 12 months at +4 °C

The immunofiltration method outlined was used for the rapid determination of specific IgG, IgA and IgM antibodies to influenza viruses A and B, RSV, Epstein-Barr virus (EBV), adenovirus, viruses causing mumps, measles, and rubella, hepatitis B virus, *Treponema pallidum, Toxoplasma gondii,* diphtheria and tetanus toxins, as well as viral and microbial antigens in sera. In most cases we obtained acceptable agreement in terms of sensitivity and specificity with the results of reference ELISA kits.

The immunofiltration method in microcolumn format was also used for sample enrichment in preparation for carrying out PCR. For this purpose we used MAb to HBsAg (Hepatitis B virus surface antigen) immobilized on the solid phase and applied 5 ml of donor plasma to the column. The sorbent was then thoroughly washed and virus DNA eluted under degrading conditions. As a result, we managed to increase the sensitivity of hepatitis B virus detection by PCR at least 10-fold.

4. Immunosensors based on atomic force microscopy and magneto-resistant technologies

One of the most promising and widely used methods for visualization of single molecules and cells, as well as the detection and localization of individual molecular interaction events, is atomic force microscopy (AFM). With the use of this method over the past 10 years, measurements have been made of the interaction forces between complementary DNA fragments, different antigen-antibody pairs, and avidin-biotin molecules [4, 5]. In addition, laboratory prototypes of detection systems using antibody affinity coatings based on AFM have been developed for protein molecules, viral particles and bacteria [6]. Nevertheless, a number of evident drawbacks of AFM as a detector element of biosensor devices (the apparatus is cumbersome and too sophisticated for effective use in serial tests) stimulated the search for new detection methods which could be adapted to rapid, serial measurements of analytes in low concentration.

The first approach is connected with the use of piezoelectric cantilever and magnetic particles with appropriate antibodies immobilized on their surface. The capacity of this method was demonstrated using the determination of proteins, viruses and bacteria as examples. The laboratory prototype of the device has been constructed.

The second approach is based on the use of magnetoresistant technologies and magnetic particles of micron dimension [7]. According to this method, magnetic beads spread on the surface of a magnetoresistant material (MRM) change the electric conductivity of the MRM considerably. The degree of magnetic particle influence on the conductivity is so high that this allows detection of the presence of only one particle on the surface of the MRM. The high effectiveness of magnetoresistant technologies, which are widely used for the creation of memory devices in modern computers, permits the formation of several thousand units of sensitive elements on 1 cm². Such a high density of detecting elements provides considerable opportunity for the analysis of multi-component mixtures as well as for multi-factor analysis of separate analytes. Trends within immunoassay development today are directed toward the simultaneous use of a large variety of antibodies with different affinities aimed at increasing specificity in the analysis of antigen and multi-component mixtures, with further information processing by computer programs and networks.

5. Conclusion

The diagnostic potential of antibody based methods for the determination of BTW agents is very high. Antibody is a unique molecule created by nature that can recognize and detect pathogens in a rapid and sensitive manner. Antibodies are important because they can be used in prophylaxis and therapy of infectious diseases. Antibody-based methods should be used in research, in investigation of pathogenesis mechanisms and in routine clinical tests. This is especially true for serology analysis and for microbial toxin detection.

6.   References

1.   Chan, D.W. (1992) *Immunoassay automatio. A practical guide*, Academic Press, Inc. Harcourt Brace Jovanovich, Publishers, San Diego, California.
2.   Köhler, G. and Milstein, C. (1975) Continuous cultures of fused cells secreting antibody of predefined specificity, *Nature* **256**, 495-498.
3.   Alakhov, V.Yu., Klinsky, E.Yu., Kolosov, M.I., Maurer-Fogy, I., Moskaleva, E. Yu., Sveshnikov, P.G., Pozdnyakova, L.P., Shemchukova, O.B., and Severin, E.S. (1992) Identification of functionally active fragments of staphylococcal enterotoxin B, *Eur. J. Biochemistry* **209**, 823-828.
4.   Lee, G.U., Chrisey, L.A., and Colton, R.J. (1994) Direct measurement of the forces between complementary strands of DNA, *Science* **266**, 771-773.
5.   Hinterdorfer, P., Baumgartner, W., Gruber, H.J., Schilcher, K., and Schindler, H. (1996) Detection and localisation of individual antibody-antigen recognition events by atomic force microscopy, *Proc. Natl. Acad. Sci. USA* **93**, 3477-3481.
6.   Babitskaya, J.I., Budashov, I.A., Chernov, S.F., Kurochkin I.N., Doroshenko, N.V., and Zubov, S.V. (1997) Langmuir films from antibodies based on amphiphilic polyelectrolytes, *Membr. Cell Biol.* **10**, 689-697.
7.   Baselt, D.R., Lee, G.U., Natesan, M., Metzger, S.W., Sheeham, P.E., and Colton, R.J. (1998) A biosensor based on magnetoresistance technology, Biosensors and Bioelectronics, *Biosensors and Bioelectronics* **13**, 731-739.

diseases. Antibody-based methods should be used in research, in investigation of pathogenesis mechanisms and in routine clinical tests. This is especially true for serotyping and for microbiota/virus detection.

References

THE POTENTIAL OF DIFFERENT BIOTECHNOLOGY METHODS IN BTW AGENT DETECTION

DNA Methods: Gene Probes

DARIO LESLIE
Defence Evaluation and Research Agency
CBD Porton Down, Salisbury, Wiltshire SP40JQ
United Kingdom

1. Introduction

The 1991 Gulf War and recent disclosures about the BW programs of Iraq and the former Soviet Union have highlighted the need for an effective biological warfare (BW) defense capability. Ideally this should include battlefield detectors as part of an integrated system with physical protection and medical countermeasures, together with a strengthened Biological and Toxin Weapons Convention (BTWC) to reduce the risk of BW proliferation.

BW agents can be either organisms or their products. Most BW agents are microbes or their toxic products and in the interests of simplicity, this chapter will discuss the use of genetic techniques to detect microbes which are either BW agents in their own right, or whose products are BW agents.

It is probable that many candidate BW agents are selected on the basis of their suitability for use in battlefield scenarios as they have traditionally been seen as battlefield weapons. BW agents are most likely to be delivered as aerosols on the battlefield and in order to maximize their effect they will have a range of shared characteristics that reflect this concept of use. These shared characteristics can be used to define some of the parameters of battlefield BW detectors. Candidate pathogens would ideally be highly infectious via the respiratory tract. This means that if they are to be detected before personnel are infected, the detector must be rapid and highly sensitive. Some agents have infectious doses of less than 10 organisms so detection of less than 1000 microbes in minutes is probably essential, and higher sensitivity is desirable. Candidate agents would ideally be stable to aerosolisation possibly via the formation of resistant spores, which may complicate detection as detectors might need to disrupt spores before identification could occur. Finally, agents are preferentially selected from microbes that are uncommon in a public health context so

A. Kelle et al. (eds.), The Role of Biotechnology in Countering BTW Agents, 79–91.
© 2001 *Kluwer Academic Publishers. Printed in the Netherlands.*

that personnel are unlikely to be immune through prior exposure or vaccination. They are probably obscure microbes that are relatively poorly studied, so reagents and methods for detection are unlikely to be available in civilian biomedicine.

On the battlefield, detection of BW agents before personnel are exposed is one of the first stages of biological defense. Thus the battlefield detector must be rapid, highly sensitive and stocked with custom reagents. In other words it will need to perform as well as or better than current laboratory equipment but must also be small, portable and automatic.

The battlefield is not the only scenario where detection of BW agents may be important. Research programs to develop offensive biological warfare capabilities have been uncovered e.g. in the former Soviet Union or by UNSCOM in Iraq and are believed to be proliferating world-wide, especially among smaller nations in politically unstable regions. One possible reason for the BTWC failing to prevent the proliferation of BW programs is that, unlike its counterpart the Chemical Weapons Convention, it contains no effective provisions for verification. International negotiations are already underway to draft proposals for a protocol to strengthen the BTWC. On-site investigation has been identified as one essential measure in the protocol, and sampling and analysis is considered to be a potentially important tool for use during such investigations [1].

Research is currently underway in various laboratories to examine technologies to address specific concerns associated with the analysis of samples collected during on-site investigations. There is more leeway in developing detectors for on-site use by investigation teams; for example, although it is desirable that technologies developed allow sample analysis within the time-frame of an on-site investigation, constraints on speed are not so severe as on the battlefield. However, the need for more detailed analysis of complex samples yielding forensic grade results without infringing upon intellectual property rights imposes significant problems.

2. Methods for Detecting Microbes

Microbes are, by definition, small and this makes them difficult to detect and identify. They are too small to be seen with the naked eye and difficult to characterize even when directly observed by microscopy. Traditionally, these problems have been circumvented by using assays to detect the presence of key components such as proteins from which the microbe's identity can be inferred. If the identification relies on the detection of components rather than analysis of the whole organism, then the choice of the components targeted is critical for success. Ideally, the target should be invariant and contain characteristic information so that it always allows identification of the organism. A number of classes of microbial components have been used for detection targets via biochemical or immunological assays, but the advances in genetics technology over the last few decades offer perhaps the best potential target of all - genetic material.

Genetic material as the *de facto* information bearer of the microbe is an obvious choice as target and contains highly stable components of great value in identification. All other microbial components are produced and controlled by the genetic material and, in informational terms, are merely second or third generation offspring of it. Genetic material is also a durable chemical that persists and retains its information-bearing capacity in a wide range of adverse conditions. Proteins, which are the main target of many other detection techniques, are less suitable because they are less informative, inherently less stable, and may not be present throughout an organism's lifecycle. For example, the bacterium *Yersinia pestis*, the causative agent of bubonic and pneumonic plague, switches the expression of many proteins on and off in response to environmental conditions such as temperature and calcium concentration [2].

Unfortunately, the ability to manipulate and analyze genetic material began decades after the equivalent eras of biochemistry and immunology that allow detection via protein markers. Although the field has developed rapidly it is still comparatively immature. Simple biochemical and immunological assay kits are common in biomedical service industries and even in the domestic retail market in the form of, for example, pregnancy test kits, while genetic assays are still almost exclusively the reserve of the specialist. However, recent advances in genetics promise significant developments in the field of genetic assays. It is widely believed that during the next decade genetic assays will approach the levels of simplicity and portability that immunological assays such as domestic pregnancy test kits offer today.

3. Genetic Material

Genetic material consists of nucleic acids (NAs) which are thread-like macromolecules made up of strands of either deoxyribonucleotides (Deoxyribonucleic acid = DNA) or ribonucleotides (Ribonucleic acid = RNA). A single strand consists of a backbone of sugar phosphates, each linked to one of four nucleotide bases. Double-stranded molecules of NA can form as a result of hydrogen bonding between bases of different strands but only certain base pairs can stably bind, so the formation of double strands depends on the two strands having the correct sequence of bases.

Both DNA and RNA can be either single or double-stranded, and hybrid double-stranded RNA/DNA duplexes can occur, but the usual forms are double-stranded DNA or single-stranded RNA. The stability of duplexes can be altered by adjusting the temperature and/or salt concentration of their environment. In this way duplexes can be destabilized to form single strands (denaturation) and single strands can be permitted to form duplexes (hybridization or annealing).

Hereditary information resides exclusively in the nucleotide sequence of NAs. Most organisms use DNA as their hereditary storage medium, but some viruses use RNA. The nucleotide sequence can be divided into functional subunits called genes and the regions between them. The nucleotide sequence within a gene represents the genetic code and describes the structure of a functional molecule, whereas the sequences between genes are either associated with the control of gene expression or form apparently superfluous packing material. When a gene is expressed (i.e. the genetic instruction is executed) a single-stranded RNA copy of the gene sequence is made which either forms a functional molecule itself, or is used as the blueprint for the construction of a protein. Proteins are the predominating functional components in living systems. NA sequences can be experimentally determined and the sequence of the bases analyzed to deduce the existence, position and function of genes and their products.

The nature of a specific region of NA will affect its suitability for use in detection and identification of microbes and its utility in different scenarios. An example of this is the degree of conservation in the NA sequence selected, i.e. whether the sequence shows a high or low degree of variability between closely related examples. In most situations genetic targets will be sought among regions of NA that show very little variation within a species. Generally, these regions will encode cellular functions that are essential for the microbe's lifecycle and thus subject to strong evolutionary pressure to remain intact. Selection of these regions minimizes the chances that strains of the target species may evade detection due to the possession of variant sequences arising either naturally or as a result of genetic manipulation. However, in situations where the identification of a certain strain of a species is required, genetic targets would be sought among hypervariable regions of nucleic acid on the basis that this may allow the differentiation of strains.

In selecting a region of NA as a target, one must also consider its function, as this will affect the information produced by the assay. At the simplest level, the information from a region of NA can be either taxonomic (i.e. what strain or species is the microbe) or functional (i.e. what does the NA encode, and by extension what can the microbe do). Functional information would ideally be related to a key attribute of the microbe. For example, in the case of a candidate BW agent, ideal functional information might be related to its virulence.

All these aspects of the NA structure and function must be taken into account in the selection of a target for a genetic assay. A full consideration of the genetics of the target microbe and the concept of use will, in most cases, lead to the selection of genetic target(s) that will provide all the necessary information in the appropriate scenarios.

4. Hybridization

Hybridization is the formation of a double-stranded molecule of NA from two single strands, and forms the basis of methods for specifically detecting NAs. Stable duplexes will only form if the two strands are completely or almost completely "complementary", i.e. they contain at every position in their sequence bases which can stably pair. On the basis of this requirement, successful hybridization is generally taken as proof, within certain confidence limits, of the identification of a NA sequence. This is because it indicates that one strand of DNA, generally termed the probe (hence gene probe), has found its complementary strand, generally termed the target, thus confirming the nucleotide sequence and identity of the target.

In traditional methods, the NA in the sample is denatured so that it is single-stranded, and immobilized onto a solid support. A gene probe, in this case an isotopically labeled single-stranded fragment of NA that is complementary to the target NA, is added to the immobilized NA. Under the correct conditions, specific hybridization occurs and the probe will bind to the target sequence if it is present. After multiple washing stages, the bound probe and hence the target sequence are detected via the probe's radioactive label.

Although twenty years have passed since the most common method of performing hybridization was first described [3], the basic procedure has changed little. It takes at least a day and requires both laboratory facilities and trained personnel. At its best it is able to detect as little as 1pg of target DNA which makes it useful as an analytical tool. However, this amount is equivalent to 10^6 target genome copies or microbes, so the sensitivity of the method is insufficient for BW detection [4]. A number of techniques have been applied to enhancing and modernizing hybridization assays, but most of these are aimed at the laboratory market and focus on simplifying the immobilization process or replacing radioisotopes with safer chromogenic labels.

In the field of biosensor research, more original approaches are being taken to radically simplifying hybridization assays. In an inversion of the traditional procedure, the gene probe is immobilized onto a sensor surface and the sample NA is passed over the immobilized probe where the target is captured. Using a fiber optic wave-guide biosensor, target binding can be monitored by measuring the accumulation of fluorescence, providing the target DNA has been fluorescently tagged [5]. Accumulation of fluorescently tagged DNA has also been reported on microchip surfaces [6] where it is detected by fluorescent microscopy. Even more promising, because they can detect the binding of unlabelled DNA, are surface plasmon resonance sensors that capture NA with probes immobilized on a metal film whose optical characteristics are altered during the binding [7].

The application of these techniques to genetic assays is in its infancy, but the potential for developing small, automated sensors is good. All interactions occur on one surface which need only be tens of microns in diameter. Sample delivery, washing and detection can be performed on that surface using a simple flow through system. Unfortunately, while sensitivity levels are poorly defined at present they seem unlikely to be better than traditional hybridization methods. Thus, even if the application of biosensor technologies allows these assays to move from the laboratory to the field over the next decade, sensitivity may still be a problem. Efforts to overcome this problem currently rely on the use of some kind of amplification process.

5. Amplification

The term amplification in this context is used to describe any process that significantly increases the amount of a reaction component. The aim of amplification is to increase the performance of a detection system to the required sensitivity levels. There are two categories of amplification: signal amplification and target amplification.

Signal amplification, where hybridization occurs and each label generates multiple signals, is inherently a part of all hybridization protocols as most procedures attach multiple labels to each probe molecule. The signal of radioisotopes can be further amplified five to tenfold by the use of intensifying screens during autoradiography. Chromogenic systems, which have been developed to replace isotopes, have enzymatic cascades where an enzyme label generates multiple products, each of which forms the substrate for the next enzyme, thus giving exponential signal amplification. Theoretically, these systems should give much better performance because of this exponential amplification, and detection of as little as 10^{-20} M analyte (approximately a thousand targets) has been reported [8]; however, achieving this type of sensitivity in assays is difficult. When enzymatic cascades are pushed to their limits inadequacies in the purification of enzymes or substrates may become significant, compromising performance and making procedures long and complex. Hence, many researchers still resort to isotopic methods when seeking unequivocal results in high sensitivity hybridizations. However, promising novel signal amplification cascades are constantly under development. For example, at CBD Porton Down a cascade using adenylate kinase to produce ATP which generates light in the presence of a genetically modified luciferase expressed from the bacterium *Escherichia coli* is currently under development. This allows the detection of 10^{-20} M of analyte and is currently being coupled to genetic assays for assessment [9].

Target amplification is the process whereby multiple copies of the target NA are made before hybridization, thus bringing the target concentration up to the working levels of the detection system. The possibility of this arose with the development of a process called the polymerase chain reaction (PCR) [10], which has revolutionized many aspects of genetic analysis [11]. In PCR, a DNA polymerase enzyme and two probes, one binding on either side of the target sequence, are added to a DNA sample. A single cycle of the PCR process is shown schematically in Figure 1. Target DNA is duplicated during a process of temperature cycling. In a single cycle of a typical PCR, DNA strands separate at 95°C, gene probe primers bind to the target DNA at 55°C, and new DNA is produced at 72°C as DNA is copied by a thermostable polymerase. Each cycle produces two copies of the target region from each single original copy and so doubles the amount of target DNA. When this is repeated many times, the target DNA is amplified exponentially and after thirty cycles a single gene may have been copied a billion times, bringing it easily within the sensitivity range of hybridization techniques.

Using PCR the detection of single gene copies is realistic, even routine in an experienced laboratory [12]. But if PCR overcomes at a stroke the sensitivity problem of hybridization assays, it does not initially overcome others. Like hybridization, PCR is a technique that in its basic form takes hours and requires laboratory staff and facilities. However, most of the time in a PCR is spent altering the temperature of large sample holders of high thermal mass, while the actual biological process takes only minutes in total. Improvement of the equipment design, for example by reducing the thermal mass of reaction vessels through microfabrication, has allowed reaction times to be reduced to minutes [13] and may soon reach the biological rather than the physical limits.

Although there are a number of NA amplification techniques that are analogous to PCR, it is the one that has been most widely accepted and characterized and so is the obvious choice for a BW detection system. Assuming that this is the case, what are the characteristics of a PCR based detection system?

6. Biological Characteristics

PCR is a technique with a huge degree of biochemical flexibility. Primers are designed by scientists and chemically synthesized, rather than being prepared in a biological system like antibodies. They can be designed to be highly specific for individual species, or generic for a group of species as has been demonstrated at CBD Porton Down with generic primers for the detection of flaviviruses [14]. They can even be designed to allow subtyping of strains of bacteria in a way analogous to forensic DNA fingerprinting of humans [15]. Samples that are analyzed by PCR do not need to be viable or even well preserved: the novel field of molecular archaeology has arisen almost solely because PCR is able to amplify NAs from samples millions of years old [16]. However, samples do need to be relatively clean if complex, time-consuming extraction procedures are to be avoided and this is an area of ongoing study both at CBD and elsewhere [17].

7. Physical Characteristics

If PCR is carried out on microfabricated devices then the physical aspects of performance are good. Hand-held battery powered devices that complete amplification in less than ten minutes have been described [13]. At this speed product detection becomes a potential bottleneck, but even here early results are promising. Hybridization biosensors are likely to take only minutes to detect the level of product generated by PCR. Analysis by microfabricated capillary electrophoresis is also possible in this time-span, although this is less desirable as product size rather than nucleotide sequence is analyzed. Possibly the most promising approach to emerge over the last few years is the monitoring of fluorescence during PCR [18]. This approach is extremely elegant as it allows the progress of PCRs to be monitored in the reaction vessel during the thermal cycling. This simplifies sample handling and reduces the risk of cross-contamination of new reactions with PCR product from completed reactions. A number of highly informative techniques are available, many based on fluorescence resonance energy transfer (FRET), which effectively senses the proximity of fluorescent markers to each other. Some of these techniques such as molecular beacons, hybridization probes and hydrolysis probes use FRET techniques to confirm, by hybridization, the nature of the PCR product as it accumulates [19, 20, 21]. Many of these techniques yield quantitative results, which allow the amount of target in the original sample to be deduced. Using rapid thermal cyclers with integrated fluorimeters, it is possible to perform PCRs and obtain an answer by FRET analysis in 10 to 15 minutes that is better than those yielded in hours by traditional PCR techniques [22].

8. Areas of Complexity

There are some areas where basic PCR methods will have to be augmented to allow the technique to become universally applicable to the detection of BW microbes. For example, PCR is a technique that works best on DNA, but some viral BW agents have RNA genomes. PCR can be applied to RNA but the assays may be more complex or less efficient. PCR assays for RNA viruses are available [14] and are based on a reverse transcription step prior to actual PCR - hence they are known as RT-PCR. Reverse transcription is the process whereby a copy of DNA is produced from an RNA template. In this case, the copy DNA is generated from the viral RNA and acts as the template for the PCR. Another example of an area where basic PCR methods may have to be supported to allow efficient detection of BW agents is in the detection of bacterial spores. A spore is a specialized cell structure that some types of bacteria can adopt in hostile environmental conditions. It generally consists of an impermeable outer layer encasing a dormant form of the bacterium. Some spores are extremely rugged and must be disrupted if DNA is to be released for PCR. We have found that such spores can be mechanically disrupted but that this takes a certain amount of time and manipulation [23]. However, the exterior coat of the spores was found to be contaminated with significant amounts of DNA that allowed the detection of approximately 10^4 bacteria without pre-processing.

9. Suitability of PCR for BW Detection

The main requirements of a genetic detection system for BW agents were discussed earlier. Essentially for the battlefield it must be fast and portable. For BTWC investigations it could be slower but might need to perform a more comprehensive analysis without infringing upon intellectual property rights. PCR based assays seem likely to meet most of these requirements in the near future. Research instruments are already almost fast and small enough to be attractive as prototype battlefield detectors; they are nowhere near being rugged or reliable enough yet, but that will come with development. Forensic laboratories already perform on human NA the types of PCR assays that would ideally be applied to BTWC investigations. Parallel techniques for bacteria and viruses are lagging surprisingly far behind considering the comparative simplicity of microbial genetics but, again, the field only needs development.

Research programs to develop genetic tests for BW agents have been underway for several years at many laboratories around the world. Many of the biological reagents and parameters to allow the detection of the candidate agents have already been developed. In their current format they have been successfully applied to the analysis of samples genuinely suspected of containing BW agents [24]. All of this work has served as proof of principle within the defense research community that genetic techniques, and especially PCR, are as attractive a proposition for BW detection as they are for biomedical detection in the civilian sectors. Research in the future should focus on developing suitable assays and equipment to allow genetic assays to be moved out of the laboratory and into the theaters where they are required. In the civilian sector there is a great deal of interest in developing and automating rapid genetic assays for clinical diagnosis. Some of this will be transferable into the BW arena, but many of the needs will be different and will require novel research to allow genetic tests to be incorporated alongside other detection techniques into suitable instruments for battlefield or BTWC investigation applications.

10. Concluding Remarks

This chapter has reviewed the field of genetic detection of microbes and its applicability to the detection of BW agents. Although it is generally accepted that genetics lags well behind other technical disciplines such as immunology in the development of simple, portable assays, there are a number of compelling reasons for pursuing genetic assays. These relate to the function of genetic material as the primary information-bearing component of living things, and the associated benefits of its stability and universal occurrence.

Many of the technical disadvantages of genetic assays are removed or reduced if a genetic amplification method such as PCR is adopted for assays. This is so fundamental and beneficial that for much of this chapter the assumption has been made that genetic assays would be PCR-based. If that is the case then it seems likely that genetic assays, in combination with appropriate confirmatory assays, will fulfill most of the requirements of the international community for the detection of BW agents in both battlefield and BTWC related scenarios.

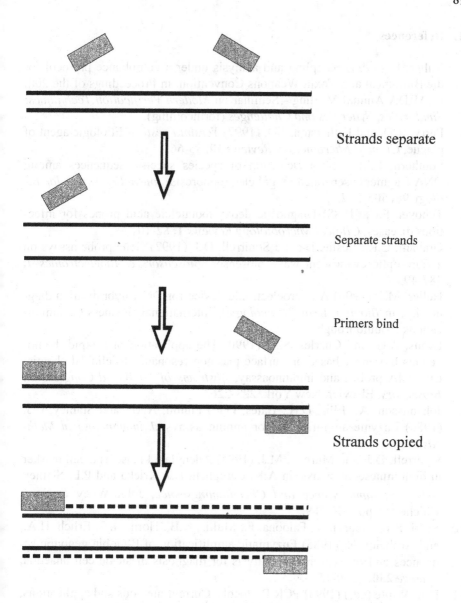

Strands separate

Separate strands

Primers bind

Strands copied

Figure 1. Polymerase Chain Reaction (PCR)
A schematic of the processes occurring within a single cycle of PCR.

11. References

1. Miller, L. (1998) Sampling and analysis under a compliance protocol for the Biological and Toxin Weapons Convention, in Proceedings of the 20th ESARDA Annual Meeting - Seminar on *Modern Verification Techniques: Similarities, Synergies and Challenges* (forthcoming).
2. Perry, R.D. and Fetherston, J.D. (1997) *Yersinia pestis* – Etiologic agent of plague, *Clinical Microbiology Reviews* **10**, 35-66.
3. Southern, E.M. (1974) Detection of species specific sequences among DNA fragments separated by gel electrophoresis, *Journal of Molecular Biology* **98**, 505-517.
4. Tenover, F.C. (1988) Diagnostic deoxyribonucleic acid probes for infectious diseases, *Clinical Microbiology Reviews* **1**, 82-101.
5. Graham, C.R., Leslie, D., and Squirrell, D.J. (1992) Gene probe assays on a fibre-optic evanescent wave biosensor, *Biosensors & Bioelectronics* **7**, 487-493.
6. Heller, M.J. (1996) A microelectronic device for DNA hybridisation diagnostics, in *Biochip Array Technologies*, International Business Communications, Southborough.
7. Evans, A.G. and Charles, S.A. (1990) The application of a rapid, homogenous biosensor based on surface plasmon resonance to clinical chemistry, DNA probes and immunoassay, *Abstracts of 1st World Congress on Biosensors*, Elsevier, New York, 223-226.
8. Johannsson, A., Ellis, D.H., Bates, D.L., Plumb, A.M., and Stanley, C.J. (1986) Enzyme amplification for immunoassays, *J. Immunological Methods* **87**, 7-11.
9. Squirrell, D.J. and Murphy, M.J. (1994) Adenylate kinase as a cell marker in bioluminescent assays, in A.K. Campbell, L.J. Kricka and P.E. Stanley (eds.), *Bioluminescence and Chemiluminescence*, John Wiley & Sons, Chichester, pp. 486-489.
10. Saiki, R.K., Scharf, S., Faloona, F., Mullis, K.B., Horn, G.T., Erlich, H.A., and Arnheim, N. (1985) Enzymatic amplification of β-globin genomic sequences and restriction site analysis for diagnosis of sickle cell anaemia, *Science* **230**, 1350-1354.
11. B.A. White (ed.) (1993) PCR Protocols. Current methods and applications, in *Methods in Microbiology* **15**, Academic Press, Harcourt Brace and Company, Ltd., London.
12. Fulop, M., Leslie, D., and Titball, R. (1996) A rapid, highly sensitive method for the detection of *Francisella tularensis* in clinical samples using the polymerase chain reaction, *American J. Tropical Medicine and Hygiene* **54**, 364-366.
13. Northrup, M.A. (1996) Advantages afforded by DNA instruments miniaturisation, in *Biochip Array Technologies*, International Business Communications, Southborough.

14. Fulop, L., Barrett, A.D.T., Phillpotts, R., Martin, K., Leslie, D., and Titball, R.W. (1993) Rapid identification of flaviviruses based on conserved NS5 sequences, *J. Virological Methods* **44**, 179-188.

15. Waterhouse, R.N. and Glover, L.A. (1993) Identification of prokaryotic repetitive DNA suitable for use as fingerprinting probes, *Applied and Environmental Microbiology* **59**, 1391-1397.

16. Pääbo, S., Higuchi, R.G., and Wilson, A.C. (1989) Ancient DNA and the polymerase chain reaction.The emerging field of molecular archaeology, *J. Biological Chemistry* **264**, 9709-9712.

17. McGregor, D.P., Forster, S., Steven, J., Adair, J., Leary, S.E.C., Leslie, D.L., Harris, W.J., and Titball, R.W. (1996) Simultaneous detection of microorganisms in soil extract based on PCR amplification of bacterial 16S rRNA fragments, *Biotechniques* **21**, 436-466.

18. Wittwer, C.T., Herrman, M.G., Moss, A.A., and Rasmussen, R.P. (1997) Continuous fluorescence monitoring of rapid cycle DNA amplification, *Biotechniques* **22**, 130-131.

19. Tyagi, S. and Kramer, F.R. (1996) Molecular beacons: probes that fluoresce upon hybridization, *Nature Biotechnology* **14**, 303-308.

20. Cardullo, R.A., Agrawal, S., Flores, C., Zamecnik, P.C., and Wolf, D.E. (1988) Detection of nucleic acid hybridization by nonradioactive fluorescence resonance energy transfer, *Proceedings of the National Academy of Sciences, USA* **85**, 8790-8794.

21. Bassam, B.J., Allen, T., Flood, S., Stevens, J., Wyatt, P., and Livak, K.J. (1996) Nucleic acid sequence detection systems: revolutionary automation for monitoring and reporting PCR products, *Australasian Biotechnology* **6**, 285-294.

22. Wittwer, C.T., Ririe, K.M., Andrew, R.V., David, D.A., Gundry, R.A., and Balis, U.J. (1997) The Lightcycler™ a microvolume multisample fluorimeter with rapid temperature control, *Biotechniques* **22**, 176-181.

23. Johns, M., Harrington, L., Titball, R.W., and Leslie, D.L. (1994) Improved methods for the detection of *Bacillus anthracis* spores by the polymerase chain reaction, *Letters in Applied Microbiology* **18**, 236-238.

24. Jackson, P.J., Hugh-Jones, M.E., Adair, D.M., Green, G., Hill, K.K., Kuske, C.R., Grinberg, L.M., Abramova, F.A., and Keim, P. (1998) PCR analysis of tissue samples from the 1979 Sverdlovsk anthrax victims: the presence of multiple *Bacillus anthracis* strains in different victims, *Proceedings of the National Academy of Sciences, USA* **95**, 1224-1229.

THE ROLE OF BIOTECHNOLOGY IN BTW AGENT DETECTION

DNA Methods: Polymerase Chain Reaction

KATHRYN NIXDORFF
Department of Microbiology and Genetics
Darmstadt University of Technology
Schnittspahnstr. 10
D-64287 Darmstadt
Germany

ALEXEI G. PRILIPOV
D.I. Ivanovsky Virology Institute
Gamaleya Str. 16
123098 Moscow, Russia

1. Introduction

In recent years it has become apparent that traditional techniques of typing bacterial strains, i.e., antibiogramm analysis, biotyping, bacteriophage typing, and serotyping, often do not provide enough discriminatory power to definitively characterize clusters of isolates that may represent an outbreak of microbial disease. At the same time, the sensitivity of traditional methods is very often too low to detect trace amounts of biological material. It is obvious that a sensitive method which can detect and unequivocally identify biological samples in a short period of time would be preferable and relevant for the detection of biological and toxin warfare (BTW) agents.

2. The Polymerase Chain Reaction (PCR)

The polymerase chain reaction (PCR) was developed in the mid 1980s and has since revolutionized molecular biology and genetics. Essentially, it is a method for the rapid amplification of DNA in vitro. DNA molecules can be multiplied up to a billionfold in the test tube in the span of a few hours by this method, yielding sufficient amounts for cloning, sequencing, or for use in mutation studies [1-4].

A. Kelle et al. (eds.), The Role of Biotechnology in Countering BTW Agents, 93–103.
© 2001 *Kluwer Academic Publishers. Printed in the Netherlands.*

PCR amplification of DNA occurs by cycles of three temperature-dependent steps: DNA denaturation, primer-template annealing, and DNA synthesis by a thermostable DNA polymerase [2-4]. In the first step, the double-stranded template DNA is denatured to single-stranded DNA. In the second step, oligonucleotide primers complementary to the 3′ends of opposite strands of the DNA are hybridized with the single-stranded template DNA. In the third step, DNA polymerase extends each primer, duplicating the segment of DNA between the two primers and completing the first cycle of denaturation, annealing and primer extension. DNA synthesized during this first cycle has the 5′end of the primers and a 3′end of variable nucleotide length, depending on how far the primer is extended before the cycle has ended. When these strands are denatured, the parental strand will rehybridize to the primer, and the product with the variable 3′end will continue to be synthesized during subsequent cycles of PCR. These products accumulate arithmetically. The second cycle of denaturation, annealing and primer extension produces discrete products with the 5′end of one primer and the 3′end of the other primer. Each strand of this discrete product is complementary to one of the two primers and thus acts as a template in subsequent cycles. Therefore, these discrete products accumulate exponentially with each following round of DNA amplification, resulting in 2^n-fold amplification of the DNA in n cycles of PCR. PCR itself has to be followed by analysis of the resulting fragment (usually by electrophoresis in either agarose or acrylamide gel, to compare the length of the resulting fragment with defined markers) in order to conclusively identify the PCR product and confirm the presence of the target sequence in the original sample.

2.1 TYPES OF AGENTS THAT CAN BE DETECTED BY PCR

PCR involves the amplification of DNA fragments, so it can be used only for DNA or for RNA, in which case the enzyme reverse transcriptase synthesizes a DNA copy (cDNA) of a messenger RNA molecule serving as a template. The cDNA can subsequently be amplified in the PCR. This method can not be used for detection of materials other than nucleic acids, including proteins. It is also necessary to know the nucleotide sequence of the target DNA in order to choose primers and define the length of the resulting amplified DNA fragment. Therefore, the PCR can detect organisms such as bacteria and viruses with at least partially known nucleotide sequences, but it can not detect toxins, because these are proteins or substances of other molecular nature that do not contain DNA or RNA. It should be noted, however, that the PCR can detect genes coding for toxins in organisms [4, 5], and indeed this possibility has particular relevance for the detection of biological warfare agents possessing toxin genes. Furthermore, unless toxins are highly purified, some DNA from the organism may be expected to contaminate samples, so that the detection by PCR of genetic material specific for genes of protein toxins could indicate the possible presence of those toxins in the samples.

2.2 LIMITS OF PCR WITH REGARD TO AMOUNTS OF AGENTS THAT CAN BE DETECTED

The PCR can be used to amplify very small amounts of DNA present in a sample. Theoretically, it is possible to find and identify a single bacterial cell in an environmental sample, even if large numbers of other microorganisms are present [1], although in practice, the sensitivity of detection of microorganisms using PCR technology varies with the samples under study. Realistic numbers that have been reported range from 18-40 genome equivalents [5, 6].

Too few cycles of PCR amplification result in a low yield of product, but too many cycles can increase the possibility of obtaining non-specific products [2-4]. Synthesis of the specific PCR product usually continues until approximately 0.3-1 picomole of DNA fragment is accumulated and a plateau is subsequently reached. However, synthesis of non-specific products may continue beyond this point and accumulate with each subsequent cycle. Therefore, too many cycles may lead to an increase the amount and complexity of non-specific products obtained. Typically, 30 cycles is a good compromise for amplifying a sample with 10^4 to 10^5 DNA molecules.

3. Detection and Identification of Organisms in Environmental Samples

In ecology studies, the PCR has had a decided positive impact, specifically in the area of identification of microorganisms in the environment. Recent investigations using the PCR have been centered on determining the diversity of microorganisms in a particular habitat, even when these organisms are not able to be cultured [7, 8]. By choosing specific primers, microorganisms can be identified at the level of groups, genera, species, or possibly even strains, provided that there are nucleotide base sequences specifying these various categories [9]. Indeed, the PCR is one of the most powerful biotechnology tools available for the identification of organisms as well as for studying biodiversity and complex ecological systems. Although PCR technology can detect small numbers of microorganisms in a mixture as mentioned above, problems still exist in many cases and methods for concentrating specimens are frequently sought [10-12]. For example, hydroxyapatite, which absorbs many biochemical substances rather non-specifically, has been used to concentrate microorganisms from ground beef, feces, and beef carcass sample suspensions [10]. In addition, silica particles have been used to capture and concentrate nucleic acids directly from clinical specimens of EDTA-plasma in the presence of guanidine thiocyanate [11], and special filters have been employed to concentrate viruses from water samples [12].

There are also numerous documentations of problems involving inhibition of the PCR, due primarily to the sensitivity of RNA and DNA polymerases to inhibitory substances in clinical and environmental samples [11, 13-20]. Some of these interfering substances include heparin, heme, polysaccharides, polyphenols, humic substances, urea, sodium dodecyl sulfate, sodium acetate, agarose, certain metal ions and salts as well as some mainly undefined substances. The methods described for purifying samples for use in the PCR usually involve absorbing the microorganisms or their nucleic acids on a matrix as in the concentration methods noted above [10-12] followed by extensive washing to remove inhibitory substances. Other methods include extraction and precipitation with organic solvents [14, 20] or applying buoyant density gradient centrifugation to the sample [16].

A very promising method that is finding ever increasing use in concentrating and purifying environmental samples for the PCR is the immunocapture-PCR (IC-PCR) [21, 22] (Figure 1). This technique employs antibodies directed against specific antigen structures (epitopes) on the surface of microorganisms. In one frequent mode of operation, these primary antibodies are coated by various methods onto small magnetic beads. When the antibodies bind specific microorganisms, these can be "fished" out of a complex sample of organisms by applying a magnet to the beads. The beads with the bound microorganisms can thus be captured and held tightly to the side of a test tube, while the rest of the organisms and substances in the complex mixture are washed away. The microorganisms captured on the beads can then be lysed and the nucleic acids recovered for use in the PCR [23]. The success of IC-PCR depends largely on the specificity and affinity of the antibodies employed.

Thus, false negative PCR results may be obtained when analyzing environmental samples that contain natural inhibitory substances or that have been deliberately spiked with such components, and this situation may be very relevant to detection and identification of agents during BW verification procedures. In general, the success of PCR carried out on environmental samples depends mainly on the purity and nature of the DNA. To detect that a sample is giving false negative results, a positive control should be included in the assay. This is a sample containing purified DNA and primers of a standard nature that are known to give a positive result in PCR, which are added to the sample under investigation. If this DNA is not amplified during PCR, this is an indication of a false negative result due to inhibitory substances in the sample. It is also important to make DNA itself accessible for PCR reaction components (e.g. polymerase, primers), so it may be necessary to destroy different constituents of either lipid or protein envelopes of microorganisms under investigation before performing PCR.

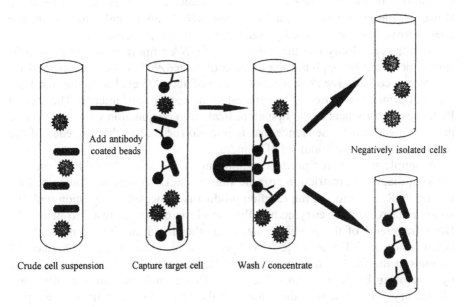

Figure 1. Concentration and selection of cells with antibodies attached to magnetic beads. Starting with a mixture of different cells, magnetic beads coated with antibodies specific for a surface structure (antigen epitope) on one of the cell types are added and allowed to react with the cells. The cell type with epitopes specific for the antibodies will bind to the magnetic beads. A magnet is then applied to the test tube, capturing the magnetic beads with the bound cells. These cells are held tightly to the wall of the test tube during washing. After removal of the cells that do not bind (negatively isolated cells), a suspension of the desired cells (positively isolated cells) can be made. Source: Ref. 23

4. Recent Innovations That Make the PCR Particularly Suitable for Detecting Agents

PCR technology is, in essence, a two step process: the amplification of a segment of nucleic acid and the detection of the resulting DNA fragment. With the rapid development of PCR technology, there has been a concomitant increase in the number of methods for PCR product detection. To be of value, a detection system must accurately and reproducibly reflect the nature and quantity of the starting nucleic acid. The simplest and most widely used detection method employs agarose gel electrophoresis to resolve and detect a product of the length predicted by the placement of the oligonucleotide primers at defined sites in the nucleotide sequence of the DNA template [3]. The appearance of a discrete band migrating in a position in the agarose gel that reflects the predited molecular size of the PCR product is usually indicative of a successful reaction. However, the presence of a band does not provide additional information, nor does it prove that the observed product is, in fact, the expected one.

Both the specificity and the sensitivity of DNA amplification can be significantly increased by applying the method of nested PCR [24]. This method involves two consecutive PCR reactions, each of about 25 cycles. In the first PCR a pair of primers is chosen to be external to the sequence of interest. The second PCR uses the products of the first amplification in conjunction with one or more primers specific for a sequence that is internal (nested) to the sequence of the region flanked by the initial set of primers.

A simple method for product monitoring is the selection of PCR primers flanking diagnostic restriction enzyme sites in the DNA region of interest. Following PCR, a portion of the reaction product is subjected to digestion with the appropriate restriction enzyme and the sample is electrophoresed to detect the DNA fragments of the predicted sizes. Another rapid monitoring method, the heteroduplex mobility assay [3], has been used for screening large numbers of samples in the field of medical genetics. Heteroduplexes are structures formed by denaturing DNA and reannealing it to DNA strands that are partially complementary. The structural distortions of the DNA double helix that are produced by such mismatched nucleotides reduce the mobility of the DNA when it is subjected to gel electrophoresis, so that it can be readily detected.

The most informative technique used to unequivocally identify a specific amplified fragment is DNA sequencing. However, sequencing is time consuming, expensive, and the analysis of a large number of samples is rather difficult. Furthermore, sequencing can not be carried out at present in a mobile laboratory.

An alternative method of identifying the PCR product short of sequencing involves the use of an oligonucleotide probe that specifically hybridizes to the template at a position located within the sequence defined by the two primers employed in the PCR. This probe can be used to detect the PCR product after its transfer to a solid support such as a nylon membrane (Southern blotting) or by solution hybridization with the product, followed by gel electrophoresis to separate the free from the bound probe [3]. Oligonucleotide probes have also proven to be essential for detecting the presence or absence of specific sequence variation.

Other approaches to the detection of PCR products include systems using a non-radioactive enzyme-linked immunosorbent assay (ELISA)-based format [3, 25]. This system utilizes biotinylated RNA probes that hybridize in solution with the PCR product. These hybrids are subsequently captured on ELISA plates that have been coated with streptavidin, which binds specifically to the biotin-labeled hybrid. In the last step, enzyme-conjugated antibodies recognizing the DNA-RNA hybrid molecules are used to quantitatively measure the amount of hybrid bound to the ELISA plate. This is determined by the strength of reaction of the enzyme with its appropriate substrate, which produces a color that can be measured spectrophotometrically. The strength of the color produced reflects the amount of hybrid formed and bound to the plate. ELISA-based formats are adaptable to numerous targets and can provide a readout compatible with quantitative PCR. Numerous other ELISA-based detection methods have been devised and, for certain molecular targets and organisms, marketed in kit form.

4.1 SUITABILITY OF PCR FOR BOTH SPECIFIC AND GENERIC DETECTION OF AGENTS

In microbiology, the basic taxonomic unit is the species. With the development of molecular genetic typing methods, the relatedness of two microorganisms is now defined largely through nucleotide sequence similarity determinations. For example, two prokaryotes (microorganisms in the domains of bacteria or archaea) whose DNA can hybridize to an extent of 70 % or greater are considered to be of the same species [1]. There are frequently genetic variants of these species that warrant their further subdivision into subspecies or strains. In the case of disease-causing agents, different strains within one species of microorganism can differ drastically with respect to pathogenic properties, so that detection of these variations along with identification of the organisms at the subspecies level can be essential in designing therapeutic strategies and establishing epidemiological links.

Multiplex PCR systems have found many applications in recent years, and have proved to be very successful in distinguishing individual variation at highly polymorphic genetic loci and identifying species- or strain-specific variations in microorganisms [3]. In these PCR systems different sites in the DNA molecule under investigation are amplified simultaneously by using several primer pairs to target different areas of the DNA. Analysis of the fragments produced can be performed by sequencing to give the most complete information. A recent extension of this method is the technique of multilocus sequence typing [26], which uses PCR amplification of a set of genetic loci coupled with automated sequencing of the products to obtain a sequence type for a particular organism, defined by the allelic profile of these genetic loci [27]. A profile based on a set of seven "housekeeping" genes, that is genes that are relatively conserved and do not change very rapidly with time, is considered useful for identifying and tracking the global spread of antibiotic-resistant bacteria and other virulent pathogens. Rapidly evolving genes on the other hand are useful for short-term, fine-scale epidemiological investigations detecting uncharacterized genomic differences between microbial isolates during an epidemic within a restricted area [26, 27].

By choosing PCR primers to amplify particular genetic loci, detection and identification of microorganisms can be made at the level of genera, species or strains. Both specific and generic detection short of sequencing can, for example, be accomplished by using a mixture of primers such as in multiplex PCR that will produce DNA fragments of defined sizes. These fragments can be analyzed in agarose electrophoresis and a conclusion as to the presence of a generic DNA fragment or specific DNA fragments can be made on-site in defined cases.

4.2 IS THE EQUIPMENT THAT IS NEEDED FOR DETECTION AND IDENTIFICATION BY PCR COMPACT AND PORTABLE ENOUGH TO BE USED IN A MOBILE LABORATORY OR VEHICLE?

The combined equipment needed for the different phases of analysis, such as DNA extraction, amplification with PCR, electrophoresis, hybridization, hybridization detection or sequencing represents a rather bulky array for a mobile on-site laboratory. However, more compact versions of instruments in use today can surely be expected in the future, because clinical laboratories and the pharmaceutical industry, which represent the primary markets for desk-top, automated equipment for nucleic acid analyses, are very interested in such developments [9]. Compact instruments that combine PCR with quantitative analysis of the reaction product have been developed and may be employed for those cases when actual sequencing of the product can be waived. In one example, a sequence-specific fluorescent signal is generated and detected in solution during PCR [28], which obviates the post-PCR steps of electrophoresis and hybridization analysis. Other high throughput, closed-tube, automated methods using fluorescent techniques of detection that combine primer and probe functions [29] are being developed.

These examples mark a trend in developments that should have a very positive impact on verification of the BTWC. The utility of PCR for on-site analysis in mobile laboratories will depend upon how well a detection system has been "formatted" to the application, that is, how well all parameters and reaction conditions in the PCR and in hybridization techniques have been standardized for a particular system in question. While it is conceivable that some analyses (such as detection of virulence genes) using systems that have been properly standardized may be carried out in appropriately equipped mobile laboratories on-site, more often than not investigations will require extensive analyses that can best be performed in better equipped off-site laboratories. Continued developments in PCR equipment towards compactness and automation can, however, be expected in the near future, so that the vision of being able to perform many PCR analyses on-site may soon become more of a reality.

5. Summary

The polymerase chain reaction (PCR) is one of the most powerful biotechnology tools available for the identification of organisms as well as for studying biodiversity and complex ecological systems. The PCR can be used to amplify very small amounts of DNA present in a sample. Theoretically, it is possible to find and identify a single bacterial cell in an environmental sample, even if large numbers of other microorganisms are present, although in practice, the sensitivity of detection of microorganisms using PCR technology varies with the samples under study. By choosing specific primers, microorganisms can be identified at the level of groups, genera, species, or possibly even strains, provided that there are nucleotide base sequences specifying these various categories. There are, however, numerous documentations of problems involving inhibition of the PCR, due primarily to the sensitivity of RNA and DNA polymerases to inhibitory substances in clinical and environmental samples. In such situations, false negative results may occur, and various methods are being tested that are designed to concentrate and purify samples from inhibitory substances. A very promising technique that is becoming wide-spread in use is the immunocapture-PCR (IC-PCR). Multiplex PCR systems have found many applications in recent years, and have proved to be very successful in distinguishing individual variation at highly polymorphic genetic loci and identifying species- or strain-specific variations in microorganisms. A profile based on a set of seven "housekeeping" genes, that is genes that are relatively conserved and do not change very rapidly with time, is considered useful for identifying and tracking the global spread of antibiotic-resistant bacteria and other virulent pathogens. Rapidly evolving genes on the other hand are useful for short-term, fine-scale epidemiological investigations detecting uncharacterized genomic differences between microbial isolates during an epidemic within a restricted area. Continued developments in PCR equipment towards compactness and automation can be expected in the near future, so that on-site analyses in mobile laboratories will become more feasible.

6. References

1. Madigan, M.T., Martinko, J.M., and Parker, J. (1997) *Brock Biology of Microorganisms*, Eighth Edition, Prentice Hall International, Inc., New Jersey.
2. Chang, S.-Y., Shih, A., and Kwok, S. (1993) Detection of variability in natural populations of viruses by polymerase chain reaction, *Methods in Enzymology* **224**, 428-438.
3. Dieffenbach, C.W. and Dveksler, G.S. (1995) *PCR primer. A laboratory manual*, Cold Spring Harbor Laboratory Press, Cold Spring Harbor, New York.
4. Maloy, S.R., Stewart, V.J., and Taylor, R.K. (1996) *Genetic analysis of pathogenic bacteria. A laboratory manual*, Cold Spring Harbor Laboratory Press, Cold Spring Harbor, New York.

5. Fagan, P.K., Hornitzky, M.A., Bettelheim, K.A., and Djordjevic, S.P. (1999) Detection of shiga-like toxin (stx_1 and stx_2), intimin ($eaeA$), and enterohemorrhagic *Escherichia coli* (EHEC) hemolysin (EHEC $hlyA$) genes in animal feces by multiplex PCR, *Applied and Environmental Microbiology* **65**, 868-872.

6. Cohen, H.J., Mechanda, S.M., and Lin, W. (1996) PCR amplification of the *fimA* gene sequence of *Salmonella typhimurium*, a specific method for detection of *Salmonella* spp., *Applied and Environmental Microbiology* **62**, 4303-4308.

7. Amann, R.I., Ludwig, W., and Schleifer, K.-H. (1995) Phylogenetic identification and in situ detection of individual microbial cells without cultivation, *Microbial Reviews* **59**, 143-169.

8. Hugenholtz, P. and Pace, N.R. (1996) Identifying microbial diversity in the natural environment: a molecular phylogenetic approach, *Trends in Biotechnology* **14**, 190-197.

9. Holzman, D. (1995) Progress toward practical DNA-based infectious disease diagnostics, *American Society for Microbiology News* **61**, 329-330.

10. Berry, E.D. and Siragusa, G.R. (1997) Hydroxyapatite adherence as a means to concentrate bacteria, *Applied and Environmental Microbiology* **63**, 4069-4074.

11. Witt, D.J. and Kemper, M. (1999) Techniques for the evaluation of nucleic acid amplification technology performance with specimens containing interfering substances: efficacy of Boom methodology for extraction of HIV-1 RNA, *J. Virological Methods* **79**, 97-111.

12. Abbaszadegan, M., Stewart, P., and LeChevallier, M. (1999) A strategy for detection of viruses in groundwater by PCR, *Applied and Environmental Microbiology* **65**, 444-449.

13. Porteous, L.A., Watrud, L.S., and Seidler, R.J. (1997) An improved method for purifying DNA from soil for polymerase chain reaction amplification and molecular ecology applications, *Molecular Ecology* **6**, 787-791.

14. Widmer, F., Shaffer, B.T., Porteous, L.A., and Seidler, R.J. (1999) Analysis of nifH gene pool complexity in soil and litter at a douglas fir forest site in the Oregon Cascade Mountain Range, *Applied and Environmental Microbiology* **65**, 374-380.

15. Sabbioni, E., Blanch, N., Baricevic, K., and Serra, M.A. (1999) Effects of trace metal compounds on HIV-1 reverse transcriptase: an in vitro study, *Biological Trace Element Research* **68**, 107-119.

16. Lindqvist, R. (1999) Detection of *Shigella* spp. in food with a nested PCR method - sensitivity and performance compared with a conventional culture method, *J. Applied Microbiology* **86**, 971-978.

17. Furuya, D., Yagihashi, A., Endoh, T., Uehara, N., Fujii, N., Chiba, S., and Watanabe, N. (1999) Simultaneous amplification of *Bordatella* repeated insertion sequences and toxin promoter region gene by polymerase chain

reaction, *Immunopharmacology and Immunotoxicology* **21**, 55-63.

18. Warnekulasuriya, M.R., Johnson, J.D., and Holliman, R.E. (1998) Detection of *Toxoplasma gondii* in cured meats, *International J. Food Microbiology* **45**, 211-215.

19. Chan, E.L., Brandt, K., Antonishyn, N., and Horsman, G.B. (1999) Minimal inhibitory effect of male urine on detection of *Chlamydia trachomatis* by Roche Amplicor PCR, *J. Medical Microbiology* **48**, 215-218.

20. Ogunjimi, A.A. and Choudary, P.V. (1999) Adsorption of endogenous polyphenols relieves the inhibition by fruit juices and fresh produce of immuno-PCR detection of *Escherichia coli* 0157:H7, *FEMS Immunology and Medical Microbiology* **23**, 213-220.

21. Jothikumar, N., Cliver, D.O., and Mariam, T.W. (1998) Immunomagnetic capture PCR for rapid concentration and detection of hepatitis A virus from environmental samples, *Applied and Environmental Microbiology* **64**, 504-508.

22. Rudi, K., Larsen, F., and Jakobsen, K.S. (1998) Detection of toxin-producing cyanobacteria by use of paramagnetic beads for cell concentration and DNA purification, *Applied and Environmental Microbiology* **64**, 34-37.

23. DYNAL Technical Handbook. Second Edition. *Cell Separation and Protein Purification* (1996) Dynal A.S., Oslo, Norway.

24. Porter-Jordan, K., Rosenberg, E.I., Keiser, J.F., Gross, J.D., Ross, A.M., Nasim, S., and Garrett, C.T. (1990) Nested polymerase chain reaction assay for the detection of cytomegalovirus overcomes false positives caused by contamination with fragmented DNA, *J. Medical Virology* **30**, 85-91.

25. Coutlee, F., Bobo, L., Mayur, K., Yolken, R.H., and Viscidi, R.P. (1989) Immunodetection of DNA with biotinylated RNA probes: a study of reactivity of a monoclonal antibody to DNA-RNA hybrids, *Analytical Biochemistry* **181**, 96-105.

26. Maiden,. M.C.J., Bygraves, J.A., Feil, E., Morelli, G., Russell, J.E., Urwin, R., Zhang, Q., Zhou, J., Zurth, K., Caugant, D.A., Feavers, I.M., Achtman, M., and Spratt, B.G. (1998) Multilocus sequence typing: a portable approach to the identification of clones within populations of pathogenic microorganisms, *Proceedings of the National Academy of Sciences, USA* **95**, 3140-3145.

27. Enright, M.C. and Spratt, B.G. (1999) Multilocus sequence typing, *Trends in Microbiology* **7**, 482-487.

28. Chen, S., Xu, R., Yee, A., Wu, K.Y., Wang, C.-N., Read, S., and deGrandis, S.A. (1998) An automated fluorescent PCR method for detection of shiga toxin-producing *Escherichia coli* in foods, *Applied and Environmental Microbiology* **64**, 4210-4116.

29. Whitcomb, D., Theaker, J., Guy, S.P., Brown, T., and Little, S. (1999) Detection of PCR products using self-probing amplicons and fluorescence, *Nature Biotechnology* **17**, 804-807.

DETECTION OF BIOLOGICAL AGENTS AND BIOSENSOR DESIGN

JAMES J. VALDES
U.S. Army Edgewood RD&E Center
Aberdeen Proving Ground, MD 21010-5423

JAMES P. CHAMBERS
University of Texas at San Antonio
San Antonio, TX 78285-0662

1. Introduction

All living things interact with their external environment by using a variety of mechanisms to recognize critical stimuli such as nutrients or light, and to respond to these stimuli. Sensory systems must therefore perform the two key functions of detection and identification. Detection functions, in turn, require the organism to collect the stimulus, in some cases to concentrate it, and to present it to the appropriate receptor. A cascade of molecular recognition, signal transduction, and data fusion and analysis events then performs identification. The environmental stimulus may be "physical" (e.g., light, sound waves) or "chemical" (e.g., odorants); in either case, the basic sensory functions of detection and identification must be carried out by the organism.

The term "biosensor" originally referred to any analytical device that detected biological materials. Thus, a mass spectrometer being used to detect a biological toxin would be, by this definition, a biosensor. The term has since evolved to denote any device that combines biological recognition elements with electronic or optical signal transduction and analysis components to detect analytes, whether chemical or biological, in the environment. Central to the concept of biosensors are the twin strategies to use principles derived from natural sensory systems to design artificial senses which mimic natural functions, and to appropriate the natural sensory system itself for detection. This is the distinction between biomimetic (i.e., synthetic systems which are analogous to natural senses) and biotic (i.e., semi-synthetic systems which use components of natural senses). In either case, the field of biosensors captured the imagination of the world's scientists and engineers as a true example of interdisciplinary research which required the co-operation of such disparate skills as molecular biology, opto-electronics, surface science, and artificial intelligence, to name a few.

A. Kelle et al. (eds.), The Role of Biotechnology in Countering BTW Agents, 105–120.
© 2001 *Kluwer Academic Publishers. Printed in the Netherlands.*

Biosensors are devices that combine biological materials with electronic or optical microsensors, a merger that results in sensitivity to biological and chemical parameters. Biosensors have been designed to detect proteins, toxins, living cells, and specific chemical compounds, but their applications are limited because selectivity and sensitivity depend on molecular recognition, which is often inadequate. Flexibility can be introduced by the addition of receptors, biological molecules capable of specific molecular recognition. The biological recognition site permits a receptor-based biosensor to detect a compound within a complex mixture of ions and chemicals in the air, water, or multicomponent liquids. This is directly analogous to the operations of receptors as they exist *in situ*, awash in a complex environment of ions, hormones, transmitters, and other biological materials, yet responding only to their appropriate ligands. For example, a biosensor that used an antibody as its recognition element would be expected to detect the antigen or hapten (the part of an antigen that can be recognized by an antibody but that is not antigenic *per se*) to which that antibody was raised; a biosensor that used the acetylcholine receptor would be considerably broader in its response, signaling the presence of any of the large number of ligands that bind to it. Immobilization of an antibody or enzyme on a sensor generally would confer its specificity on the sensor; conversely, a receptor with a more broad-spectrum recognition capability would confer like quantities on the biosensor.

It was also readily apparent that biosensors had nearly unmatched potential for both commercial and military applications. The imagination was fired by dreams of smart materials which could sense and respond to strain, as bones do; intelligent sensors which could control manufacturing processes; detectors for environmental pollutants; medical diagnostics; and many other applications. However, the unmatched potential tended to obfuscate the very difficult technological challenges inherent in this marriage of biological and synthetic materials, and biosensors would remain laboratory curiosities for many years.

2. The "Nature of the Beast" – What to Detect?

2.1 THREAT

The singular characteristic of the biological world is competition, and Nature has produced an incredibly diverse spectrum of dangerous plants, animals, microorganisms, viruses and infectious proteins which are lethal to each other – or humans – and to which we have few early warning sentinels. This poses enormous problems for public health, as we invade the eco-niches of exotic diseases which are then spread by increasingly mobile populations; agriculture, as crops become genetically more homogeneous, hence susceptible to catastrophic loss;

military force protection and homeland defense, as the ability to produce these natural products becomes easier due to advances in biotechnology.

Biological agents derived from natural products have been developed as weapons of mass destruction and represent the primary strategic threat to democratic governments. Unlike conventional or chemical munitions, pathogens such as viruses and bacteria, and infectious proteins such as prions, are self-replicating, hence very small quantities can be used in an attack and the effect amplified by secondary infection. Add to this the potential for the use of chemical agents in low-intensity conflicts or terrorist attacks, and the recent advances in biotechnology which have blurred the distinction between biological and chemical agents, and the problem of detection begins to appear unsolvable.

2.2 NATURE OF BIOLOGICAL WEAPONS

The highly sophisticated, well funded and focused biological weapons program of the former Soviet Union was prototypic of a major power's weapons program and somewhat mirrored the U.S. program. As such, it was relatively easy to understand and its goals and doctrine were clear. Biological weapons in the hands of third tier national powers, transnational groups and even individuals represent an entirely new, and poorly understood, threat. Rather than being an offsetting capability to another major power's equally lethal arsenal, biological weapons now represent an asymmetrical response to a technologically and militarily superior adversary. The nature of the threat and doctrine for use by such untraditional adversaries are difficult to predict.

In general, it can be said that biological weapons are accessible, flexible and difficult to defend against. They are accessible because they are easy to produce and somewhat easy to disseminate, given a basic understanding of microbiology and aerosol science. Production of biological agents is relatively inexpensive and based on commercial technology; in fact, all the technology required is thoroughly dual use. They are flexible for a number of reasons. There are lethal (e.g., anthrax) and non-lethal (e.g., staph enterotoxin B) options. Their use can be tactical, as on the battlefield or against fixed sites such as airfields; strategic, as launched against cities in bombs or missiles; unconventional, as in terrorist operations. Further, the targets of biological agents include: crops, cattle and other economic targets; materiel and infrastructure (e.g., biocorrosion); personnel.

For all of the above reasons they are difficult to defend against. It is also likely that the tools of molecular genetics will yield a new generation of advanced biological weapons.

2.3 CHARACTERISTICS OF BIOLOGICAL AGENTS

Biological agents, in order to be effective as weapons, require certain characteristics. [14] The ideal biological agent is pathogenic to humans, plants or animals or deleterious to materiel. It must be producible in quantity and concentrated to yield a high percentage of agent. It must be stable, or made stable by the use of adjuvants, modifications or packaging such as microencapsulation. It must be sized to respirable sized particles (1-5 microns) and be stable during dissemination. Depending on the application, the agent must be infectious but generally not contagious, and effective at very low doses. Finally, it is generally believed that an ideal agent would have medical counter-measures available, but this is probably wishful thinking based on a western mindset. These characteristics of biological agents and weapons to a large extent determine how biosensors will be designed and employed.

3.   The Tools at Hand

3.1 SOME QUESTIONS

There are several key issues that determine the approach to designing a biosensor. The first is simply, how quickly does one need the answer? Detection of, say, a pollutant in an environmental remediation site does not demand a near real-time response from the biosensor and may, in fact, be better accomplished by more traditional analytical laboratory techniques. Conversely, soldiers on the battlefield or the policeman responding to a terrorist attack need real-time response as well as an "all clear" signal.

The second key issue is one of detection versus identification. Does one want to know that something – anything – dangerous has happened in order to make a decision on whether or not to don protective gear, or specifically what the agent was. The soldier in combat could live with the former, more rudimentary, capability, whereas the physician or the treaty inspector requires absolute identification. These functional requirements are the basis of the debate over "specific" versus "generic" detection.

A third key question is whether one has an *a priori* knowledge of what is to be detected. If the soldier knows that the adversary has weaponized anthrax, or the environmental inspector knows that the factory produces particular organic solvents, the biosensor could be designed based on the agent to be detected. If, however, this knowledge were lacking, a more target-oriented strategy would be preferable. One such strategy would be to use physiological receptors as the recognition element of the biosensor. [17] The lethal or incapacitating effects of many chemicals and toxins, the pharmacologic effects of drugs, and the action of viruses result from molecular recognition of these agents by physiological targets known, generically, as receptors. Receptors are macromolecular binding proteins that are vital for cellular function and can recognize and respond to agents at extremely low concentrations. Threat potential and pharmacologic potency are direct correlates of receptor affinity, making receptors, by nature, extremely efficient detectors. Receptor-based assays are used routinely in the laboratory, but their use as detectors has only lately been proposed. These will be discussed in more detail in a later section.

Fourth, what is the nature of what is to be detected? If it were simply a matter of detecting an increase in the number of particles in the environment, a simple particle counter would be sufficient. Detection of whole cells offers unique challenges due to capture difficulties, and differential expression of target epitopes during different phases of growth. Detection of cellular components such as genetic material usually requires more extensive sample preparation such as cellular disruption and removal of interferents, whereas detection of a particular toxin can be easily accomplished with an immunoassay and minimal sample preparation.

Finally, does one want to know if the pathogen is viable (or the toxin active) or merely present. The soldier or policemen needs to know when it is safe to remove protective gear, while the treaty inspector may simply want evidence of a violation which would not require a viable sample.

3.2 SOME CAVEATS

In a defensive world in which one has perfect knowledge of an adversary's capabilities, and can produce perfect vaccines against all these threats, there would be no need for detectors except for verification applications. However, our knowledge of the threat is imperfect and vaccines are not 100 percent effective. Since we can never absolutely predict the threat, we need broad-spectrum detectors; since we must diagnose and treat casualties, we need detectors specific enough to identify the agents; and, since we must identify agents in noncompliant situations with absolute certitude, we need flawless technology. [16] An integrated approach is therefore required.

3.3 TRADITIONAL APPROACHES AND TECHNOLOGICAL INNOVATIONS

The traditional approaches to detecting bacteria and viruses date back to Louis Pasteur and are still in use today. Samples suspected of harboring bacteria are grown in culture and the bacteria identified by a number of physical and chemical characteristics. Viruses, having no intrinsic metabolic activity and requiring another cell's metabolic machinery for replication, are grown in a suitable host cell. Finally, tissue response to toxins (i.e., bioassays) is evaluated using a number of unique tissue systems such as the perfused rat hemi-diaphragm. More recently, assays using isolated receptors and radioisotopically tagged ligands have largely replaced traditional bioassays.

Two key technological innovations have revolutionized biological detection and made biosensors far more viable and flexible. The first was the ability to generate large quantities of monoclonal antibodies (MAb) using hybridoma technology, an approach which is now being supplanted by genetic libraries and recombinant antibodies. The availability of large numbers of MAbs allows the design of biosensors for virtually any substance or organism to which a MAb can be created. The second innovation was the development of nucleic acid based techniques such as the polymerase chain reaction (PCR). This ability to amplify minute amounts of genetic material is the critical step in the design of biosensors for any pathogens for which suitable oligonucleotide probes and primers exist.

These innovations and their impact on biosensors will be discussed in more detail in later sections. In both cases, however, two considerations are extent: Both are specific to a particular biological or chemical agent, and both have certain limitations when configured in a fieldable biosensor or assay.

3.3.1 Components of a Biosensor

Biosensors are composed of a biological recognition site, whether biotic or biomimetic, a signal transduction capability, usually optical or electronic, and data fusion and analysis. The recognition elements will be described in some detail and, while the emphasis of this paper is on antibodies, nucleic acid probes and polymer imprints, other structures have been used. These include neuro-receptors, enzymes, peptide arrays, lectins and even whole cells.

Just as the particular application dictates which biological recognition site is used, it also is a key determinant in the choice of microsensor. These include, but are not limited to, planar capacitive chips, piezoelectric crystals, surface and shear horizontal acoustic wave devices, surface plasmon resonance, ampero-metric and potentiometric sensors, evanescent wave sensors and direct detection such as microcantilevers.

3.3.2 Recombinant Antibodies

The practical use of antibodies for diagnostics and therapy was made possible by the development of hybridoma technology in 1975. [9] This is a method to create immortal cell lines which produce antibodies by fusing B cells with cancer cells, the former providing the antibody production capability and the latter immortality. Hybridoma technology is currently the antibody production method of choice, but has several serious limitations: Immunizing animals is a laborious process which takes months; hybridoma cells are notoriously finicky and must be grown in expensive medium under sterile conditions; genetic drift often results in a cessation of antibody production in otherwise healthy cells; and, finally, yields are fairly low.

A bacteriophage is a virus which infects bacteria, but is harmless to humans. The phage has genes which code for a minor coat protein on its surface called cpIII, and a major coat protein called cpVIII. By fusing the gene which codes for an antibody to the coat protein gene, it is possible to make the phage express the antibody on its surface. The phage with this gene is then isolated by affinity selection using the antigen of interest, and further rounds of selection can be performed to enrich the yield a million fold or more. Thus, even when there exist only a few phage in a population of billions with the correct antibody gene, they can be isolated in relatively short order. Bacteria are then infected with the phage for rapid production of the antibody using standard fermentation technology.

It is readily apparent that this technology solves the production problem, because bacteria are cheap and easy to grow in large quantities. However, the process still begins with an immunized animal and a hybridoma cell line, and is therefore time consuming. Hybridomas have subsequently been removed from the process by cloning antibody genes directly from the B lymphocytes, inserting them into phage, and infecting bacteria as before.

It is likely, given the pace of progress of biotechnology, that a unique threat agent may be encountered to which no detection system has been designed. It is now possible to by-pass animals completely by constructing a synthetic repertoire of antibody genes, eliminating the need to immunize. This "super library" is a collection of all the billion or so possible genetic combinations in the immune system and would not have the limiting bias of an immunized animal, would eliminate the need for injection schedules and harvesting tissues, and could function as a synthetic immune system in which antibody selection occurs *in vitro.* [7] The super library could be prepared in advance and stored until needed, at which time the library would be screened for antibodies which recognize the new agent. Literally billions (trillions, including mutations) of possible antibodies could be rapidly screened and production could commence immediately. In principle, a single super library could generate human antibodies against any antigen or threat agent that the intact immune system can recognize and would therefore mimic the immune response.

In summary, the immune system is Nature's ultimate detection and protection system. The reagents which are coupled to biosensors include: Polyclonal antibodies, which represent an undefined mix of different antibodies; monoclonal antibodies, which are derived from one clone and produced in hybridomas; defined "polyclonal" mixtures, which are combinations of known monoclones; recombinant single chain (ScFv) or fragments of antigen binding (Fab), which are well defined gene products usually produced in bacterial or fungal fermentations. The advantages of the recombinant approach are (1) production of a well characterized gene product, (2) lower cost, rapid large-scale production via fermentation, (3) ability to screen large numbers of clones and obtain rare epitope binders, and (4) amenability to genetic manipulation.

3.3.3 Nucleic Acid Based Approaches

In principle, any organism, even an identical human twin, can be identified on the basis of its unique nucleic acid sequences. For lower organisms such as prokaryotes whose genomes are quickly being completely sequenced, the development of several assay formats is quite sophisticated.

All nucleic acid-based assays rely upon the ability of a oligonucleotide strand to hybridize – form a stable complex – with a complementary strand from the target organism. Most direct hybridization assays use an oligonucleotide probe which has been covalently labeled with a reporter molecule. These reporters can include radioisotopes, enzymes or fluorophores and serve to generate a detectable signal. This eliminates post-hybridization amplification such as PCR [12] but can often interfere with the ability of the probe to bind to its target. Fluorescent and chemiluminescent signals can be read on small fluorometers or spectrophotometers.

One important advantage of direct hybridization assays versus PCR is that they are more easily quantified. Their major disadvantage, lack of sensitivity, is being addressed by a new technique called Branched DNA Signal Amplification, in which sequential hybridization of sets of probes results in a 10^4 increase in signal. Sensitive detection of DNA/RNA using direct hybridization has been reported for the Light Addressable Potentiometric Sensor (LAPS), the Fiber Optic Waveguide (FOWG) and a Charged Couple Device. Detection of 20pg of DNA in a LAPS configured sandwich assay has been reported, and 24 femtomoles of DNA was detected using direct capture on a FOWG. [10]

Nucleic acid-based detection has been vastly improved by development of amplification processes which increase the number of targets. Most well known of these is the PCR, which involves the enzymatic replication of a target region of nucleic acid, defined by sets of specific oligonucleotide primers. The salient feature of PCR is that each product strand becomes a template for a subsequent round of replication and that, in theory, detection of a single genetic target should be possible.

Early PCR methodologies involved the use of *E. coli* DNA polymerase and repeated cycles consisting of three steps: denaturation, annealing and extension. The discovery and implementation of a thermostable DNA polymerase from *Thermus acquaticus* as well as the advent of programmable thermocyclers vastly improved the process, and PCR has been applied to the development of detection systems for countless bacteria and viruses. For viruses that lack a DNA genome, an initial reverse transcription step is required since *T. acquaticus* DNA polymerase (*Taq* polymerase) does not utilize RNA as a template. Other DNA polymerases, most notably *Thermus thermophilus* DNA polymerase (*Tth* DNA polymerase) have both a DNA polymerase activity as well as a reverse transcriptase activity. As a consequence, a single tube reaction can be carried out.

PCR can be coupled to virtually any nucleic acid-based detection system, resulting in a significant increase in the assay sensitivity. Therefore hybridization-based assays with an initial PCR step to increase target concentration have been successfully applied to detect a vast number of different targets. One novel system of note is the 5' nuclease assay which employs a fluorogenic probe. In addition to DNA polymerase activity, *Taq* polymerase among other thermostable polymerases, also has a 5' nucleolytic activity which hydrolyses any nucleic acid molecule that is hybridized to the template in its path. The fluorogenic probe is dual-labeled with a fluorophore reporter on the 5' end and a second fluorophore which is able to quench the emission intensity of the reporter dye. The quenching is distance dependent and the distance remains relatively constant when both fluorophores are coalently attached to the oligonucleotide. The DNA polymerase, however, will hydrolyze the fluorogenic probe if it is hybridized to the amplicon lying in the path of the enzyme. Upon hydrolysis the fluorophore reporter is released and its fluorescence intensity increases.

An early deficiency of PCR was its inability to discriminate between targets that differed in only a single base pair. An alternative amplification scheme was therefore developed called the ligase chain reaction (LCR) which was derived from the ligation detection reaction (LDR). LDR involves two adjacent oligonucleotides designed such that the junction between the 3' end of the upstream primer and the downstream primer coincides with the nucleotide that distinguishes one target from another. DNA ligase will seal the nick between these two oligonucleotides only if, the 3' end of the upstream primer is complementary to the target. Ligation of the two oligonucleotides indicates a positive reaction and occurs only when the appropriate target is present. In LCR, two sets of primers are designed whose junction overlaps the nucleotide to be interrogated. Similar to LDR, a ligated product is only produced when the target matches the two diagonally opposed oligonucleotides. In LCR the ligated product then can serve as a template for a subsequent round of ligation resulting in a two-fold increase in the number of templates.

3.3.4 Artificial Receptors

One of the crucial limitations of biosensors is that the biological reagents used for molecular recognition, be they antibodies, gene probes or enzymes, are not particularly stable in the environment. This limited shelf life and operational range is a result of the constraints posed by narrow temperature, pH ionic, and other requirements. One possible solution is the design of synthetic receptors using molecularly imprinted polymers.

The stable properties of molecular imprints make them ideal artificial receptors and a first step towards "reagentless" detection. Molecular imprints based on cross-linked silane polymers are physically stable and may be used to detect a wide range of chemical and biological molecules such as peptide toxins. Molecularly imprinted organic silanes retain both the binding activity and the biological activity of the imprinted analyte, and several methods may be used to determine imprint occupancy. These include radio-labeling, high performance liquid chromatography, liquid chromatography and direct fluorescence detection. The latter is useful with compounds which contain the aromatic amino acids tryptophan and tyrosine or some nucleic acids, or which can be derivatized with fluorescent dyes.

Fluorescence detection is advantageous because it requires small samples and offers flexible monitoring parameters such as spectral lifetime, shift, quench and anisotropy. Certain limitations of this approach must be considered; for example, fluorescent intensity may not be directly proportional to sample concentration due to excited state reactions or non-emissive decay, and ground state aggregation and complex formation may also result in inactivation of a population of fluorophores. Nevertheless, molecules exhibiting intrinsic fluorescence provide the ideal model system. Where this characteristic is absent, one may envisage imprint binding sites labeled with fluorescent probes or coupled to fluorescence–linked assays in which the binding event would quench, enhance or subtly alter fluorescence emission.

The use of a silica support for silane-based molecular imprints increases mechanical stability and suggests a myriad of applications under non-optimized field conditions. While this approach has proven successful with small and medium molecular weight toxins such as conotoxin and ricin, respectively, it has the potential to be extended to many classes of compounds. [8]

4. Selected Sensors

4.1 EVANESCENT WAVE SENSORS (EWS)

EWS designs involve the use of an optical waveguide, configured as either a fiber or a planer wafer, through which light propagates. The light source can be a quartz halogen lamp, light emitting diode or a laser, and the sensor is typically configured as in Figure 1.

Light propagates within the waveguide and, at each point of reflection, creates an electromagnetic field called the evanescent effect, which penetrates perpendicular to the waveguide into the adjacent medium. This "evanescent zone" surrounds the fiber and will allow optically tagged capture molecules to be excited, with their emitted light tunneling back into the fiber. The waveguide thus combines the functions of binding, amplification and detection. Typically, capture molecules such as antibodies are immobilized on the fiber with their fluorescinated antigen bound. If the sample being tested contains antigen, it will compete with the labeled antigen for the antibody binding site, thus altering the optical output of the EWS.

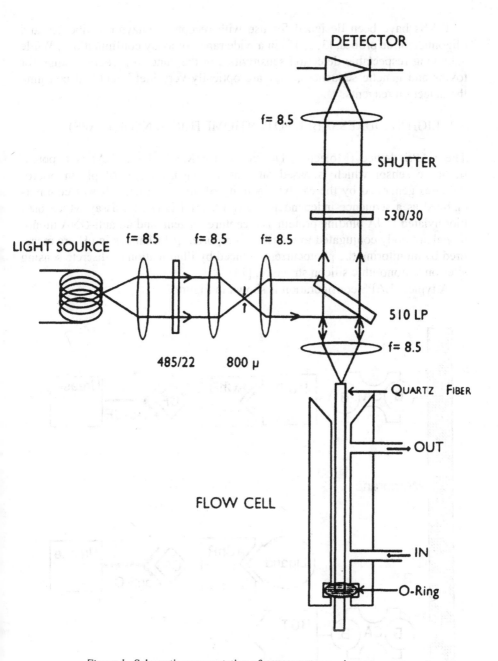

Figure 1. Schematic representation of evanescent wave immunosensor.

Source: K.R. Rogers, J.J. Valdes, and Eldefrawi, Anal.Biochem., 182 (1989), 353-359.

EWSs have been designed for use with receptors, enzyme antibodies and oligonucleotide probes, [3, 5, 15] in a wide range of assay configurations. While achieving respectable detection sensitivities in the nano to picogram range for toxins and genetic sequences, they are optically very inefficient and consume the detection reagents.

4.2 LIGHT ADDRESSABLE POTENTIOMETRIC SENSOR (LAPS)

The LAPS device (Molecular Devices Corp; Menlo Park, CA) is a potentiometric sensor which is based on sensitive measurement of pH in microvolumes generated by the catalytic activity of urease on urea. It was commercialized as a sequence-independent assay for total DNA, an assay which uses biotinylated DNA binding protein as a capture reagent, and an anti-DNA monoclonal antibody conjugated to urease as a reporter. [6] Changes in pH are measured by monitoring the photocurrent induced by illumination of discrete sensing sites on a monolithic silicon structure. [13]

A typical LAPS configuration is shown in Figure 2.

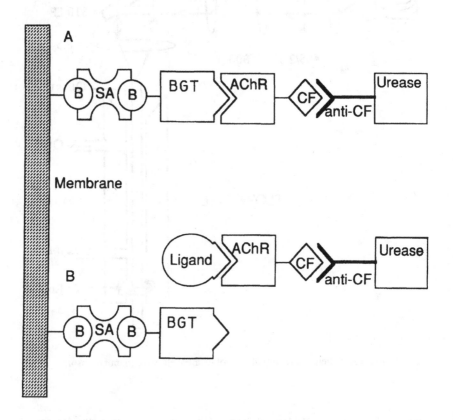

Figure 2. Configuration of the light addressable potentiometric sensor (LAPS)

A biotinylated dipstick is conjugated with streptavidin and the following reagents are used: biotinylated antibody raised against the antigen to be detected, or a biotinylated oligonucleotide probe specific for a gene sequence to be detected; fluoresceinated antibody raised against another epitope of the antigen to be detected, or fluoresceinated oligonucleotide probe raised against another part of the gene sequence; anti-fluorescein antibody conjugated with urease. If the target of interest is present in the sample, all of the above binding events will occur. The dipstick is then inserted into the reader where it is in proximity to the silicon chip, and measurements of voltage changes are made as the urease reacts with urea.

4.3 CYTOSENSOR MICROPHYSIOMETER

LAPS technology has been shown to be exquisitely sensitive for detecting and quantitating the binding of an analyte to its biologic receptor, and has now been extended to the detection of metabolic responses in living cells. A complete description of the technology is available elsewhere. [11] Briefly, cells of any type are maintained in culture in low buffered medium containing glucose, which is metabolized to lactic acid. The acidification rate of the media in the microenvironment of the LAPS device results from the cellular metabolic activity, and any change in metabolism can be detected by changes in pH of as little as 0.01 pH unites.

The cytosensor can be used to assess both the general metabolic condition of cells, as well as specific mechanistic effects of drugs, toxic industrial chemicals, toxins and chemical agents on specific receptor and ion channel subtypes. Two major advantages are that almost any cell type – e.g., human, bacterial, protozoan – can be used, and metabolic changes can be measured in real time (i.e., seconds) without any significant changes in cell morphology or physiology. [4] Thus, it represents a very powerful tool for environmental monitoring, drug discovery and pharmacologic studies at the cellular and molecular level. Correlations of IC_{50} volumes for 24 hr exposure with ten tested drugs and their published human lethal blood concentrations were excellent (r=0.958). [1]

4.4 SOLID STATE OPTICAL SENSOR

Most biosensors suffer from the parallel drawbacks of requiring reagents such as antibodies or oligonucleotide probes, and of consuming these reagents during the detection process. These reagents also require elaborate fluidic systems, resulting in a system that is difficult to harden for field use. It is therefore clear that the two key characteristics of future sensors are that they be (1) reagentless or solid state and (2) regenerable.

Recently, a solid state optical sensor platform has been described which may meet the above requirements. [2] While this device was designed for oxygen measurements, its flexibility makes it amenable for use with a wide range of biological recognition sites. The design, shown in Figure 3, uses a solid state sandwich of a GaN light emitting diode, an optical filter in proximity to a polymer doped with a fluorescent indicator, and a photodiode. By using different polymers and doping them with appropriately tagged biological recognition sites, it should be possible to detect anything for which a capture molecule is available. The twin design elements of allowing analyte diffusion in a path normal to the optical absorption path, and of putting the LED and photodiode in close proximity, result in very low power requirements, because of vastly increased optical efficiency compared to EWS systems, and the potential for extreme miniaturization.

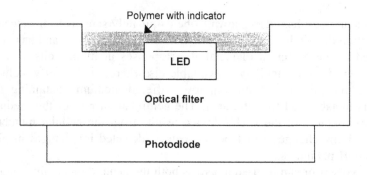

Figure 3. Solid state optical sensor

5. Conclusion

There exist a myriad of sensor platforms, biological recognition molecules, transduction mechanisms and assay configurations in numbers far beyond the scope of this chapter to describe. It is, however, the particular application that will drive the choice of technologies, and design decisions will be based on the advantages and disadvantages of the various configurations.

In general, future systems will share three characteristics. They will be essentially reagentless and solid state, both due to use and cost considerations. They will be microminiaturized, including being reduced to nanoscale arrays. Last, they will be intelligent systems, capable of learning patterns of recognition and modifying their responses in the future.

6. References

1. Patrick, W. (1995) Rediscovering biological weapons, *Chemical and Biological Warfare Proliferation Course*, Central Intelligence Agency, Chemical and Biological Warfare Branch, McLean, VA.
2. Valdes, J.J., Wall, J.G., Chambers, J.P. and Eldefrawi, M.E. (1988) A receptor-based capacitive biosensor, *Johns Hopkins APL Technical digest* 1, 4-10.
3. Valdes, J.J. (1997) Biological agent detection technology, *New Scientific and Technical Aspects of Verification of the Biological and Toxin Weapons Convention*, NATO Advanced Studies Institute Series (Graham Pearson, editor).
4. Kohler, G. and Milstein, L. (1975) Continuous cultures of fused cells secreting antibody of predefined specificity, *Nature* **256**, 495-497.
5. Hoogenboom, H.R. (1997) Designing and optimizing library selection strategies for generating high affinity antibodies, *Trends in Biotechnology* **155**, 335-350.
6. Mullis, K. and Faloona, F.A. (1987) Specific synthesis of DNA in vitro via a polymerase catalyzed chain reaction, *Methods in Enzymology* **155**, 335-350.
7. Kumar, P., Willson, R.C., Valdes, J.J., and Chambers, J.P. (1994) Monitoring of oligonucleotide hybridization using light addressable potentiometric and evanescent wave fluorescence sensing, *Materials Science and Engineering* **1**, 187-192.
8. Iqbal, S.F., Lukla, M.F., Chambers, J.P., Thompson, R.G., and Valdes, J.J. (2000) Artificial Receptors: molecular imprints discern closely related related toxins, *Materials Science and Engineering,* **7**, 77-81.
9. Eldefrawi, M.E., Eldefrawi, A.T., Wright, J., Emanuel, P.A., Valdes, J.J., and Rogers, K.R. (1995) Immunosensors for detection of chemical mixtures, *Biosensor and Chemical Sensor Technology, ACS Symposium Series 613*, **3**, 19-32.
10. Colston, J.T., Kumar, P.R., Epple, D., Tsin, A.T.C., Valdes. J.J., and Chambers, J.P. (1993) Detection of sub-nanogram quantities of mojave toxin via enzyme immunoassay with a light addressable potentiometric detector, *Biosensors and Bioelectronics* **8**, 117-121.
11. Song, X., Nolan, J., and Swanson, B.I. (1998) Optical signed transduction triggered by protein-ligan binding – detection of toxin using multivalent binding, *J. American Chemical Society* **120**, 4873-4874.
12. Hafeman, D.G., Parce, J.W., and McConnell, H.M. (1988) Light addressable potentiometric sensor for biochemical systems, *Science* **240**, 1182-1185.
13. Olsen, J.D., Panfili, P.R., Armenta, N., Fennel, M.B., Merrick, H., Gumpeiz, J., Goltz, M., and Zuk, R.F. (1990) A silicon sensor-based filtration

120

immunoassay using biotin-mediated capture, *Immunological Methods* **134**, 71-79.

14. McConnell, H.M., Owicki, J.C., Parce, J.W., Miller, D.L., Baxter, G.T., Wada, H.G., and Pitchford, S. (1992) The cytosensor microphysiometer – biological application of silicon technology, *Science* **257**, 1906-1912.

15. Eldefrawi, A.T., Cao, C.J., Cortes, V.I., Mioduszewski, R.J., Menking, D.E., and Valdes, J.J. (1996) Eukaryotic cell biosensor: the cytosensor microphysiometer, *Methods in Biotechnology: Affinity Biosensors: Techniques and Protocols* **7**, 223-238.

16. Cao, C.J., Mioduszewski, R.J., Menking, D.E., Valdes, J.J., Cortes, V.I., Eldefrawi, M.E., and Eldefrawi, A.T. (1997) Validation of the cytosensor for in vitro cytotoxicity studies, *Toxicology In Vitro* **11**, 258-293.

17. Colvin, A.E., Phillips, T.E., Miragliotta, J.A., Givens, R.B., and Bargeron, C.B. (1996) A novel solid state oxygen sensor, *Johns Hopkins APL Technical Digest* **17(4)**, 377-385.

Acknowledgements

The authors thank Mrs. Marcia M. Goforth for preparation of the manuscript and Professor Mohyee E. Eldefrawi for his years of collaboration and intellectual guidance in the field of biosensors.

DETECTING BTW AGENTS ON THE BATTLEFIELD

P.C.C. HENRI GARRIGUE
Ministry of Defence, DGA/DCE
Centre d'Etudes du Bouchet
BP 3, 91710 Vert le Petit, France

1. Introduction

Before beginning to discuss the concept of detection of biological agents and of the techniques which can be used, it is important to recall some characteristics of the biological threat. Generally, BTW agents have a non-immediate effect on people and for live agents, an incubation period is necessary. This incubation time can last from half a day to weeks or months. In the particular case of toxins, effects are obtained after hours or days. For most BTW agents recognized as belonging to the biological risk category, a therapy (antibiotic/immuno-therapy) exists, is effective during the hours immediately after exposure and diminishes clinical symptoms. For a few agents vaccines exist even if they are not widely available. Agents are numerous; detection of BTW agents and available identification techniques are time-consuming and must be carried out to give an answer as quickly and precisely as possible.

These characteristics lead to a strategy of biological defense slightly different from the strategy for chemical weapons agents:
- The necessity for rapid detection and identification followed by an immediate report of results, in order to allow specific protection measures and therapy (it can be argued that if agents need a long incubation time to affect people, and if therapy is known, speed is not that crucial; in fact, the more people that are protected the better it is, firstly for them, and secondly in minimizing the logistical problems of treatment).
- The detection and identification concept must be based on the use of a set of complementary equipment and orthogonal techniques. One of the main challenges of the strategy will be to acquire adequate specificity:
  - Specificity of detection: the current methods are based on biological detection, immunodetection and nucleic acid probes which require the use of specific reagents even if a poor generic detection system can be used. This specificity means that in the absence of the correct reagents, no detection is possible. Only what is looked for can be discovered.

121

A. Kelle et al. (eds.), The Role of Biotechnology in Countering BTW Agents, 121–128.
© 2001 *Kluwer Academic Publishers. Printed in the Netherlands.*

– Specificity of therapy: a vaccine against one agent gives no protection against other agents. In the case of bacteria, the spectrum of an antibiotic can be more or less wide but cannot cover the whole spectrum of agents, and furthermore the use of genetic modification techniques can widen the classical spectrum of resistance of agents. For viruses, specificity, apart from vaccination, is not really a consideration because virucides are not effective. For toxins, apart from vaccination, therapy is based on sera or on symptoms.

2.   Biological Detection: A Concept and a Strategy to Achieve the Goal

The concept of detection looks like the concept developed for the detection of chemical agents in its totality, but many differences exist because of the above characteristics. The classical scheme uses the following steps: alarm detection, control detection and unambiguous detection:
– Alarm detection is needed to alert the chain of command as soon as possible to the danger and allow it to take any necessary measures to protect soldiers. Its primary function is to protect soldiers early enough to avoid casualties.
– Control detection is needed to confirm (or not) the presence of agents and justify the protection measures; at the same time it must allow monitoring of the evolution of the risk for the soldiers and, as protection means a diminution of the operational capability, must permit removal of the protective measures as soon as possible when the danger is over.
– Unambiguous detection is then needed to obtain indisputable proof of the presence of the agent on the battlefield and allow some type of further political response.
In this classical scheme, something is missing - the identification of the agent. As has been shown earlier, identification is essential because it reveals the nature of the agent and allows specific therapy to treat people. When biological detection is used, in fact, identification occurs at the same time as detection because of the specificity of the system. A more precise identification may be required for specific strains and in the case of genetically modified agents.

3.   The State of the Art in Available Technologies

Techniques which can be used for the identification of agents are numerous but can be classified in three main groups. Each of them has its own advantages, limitations and costs. Actually, lessons from analysis of unknown samples show the necessity of conducting, in parallel, classical microbiology, immunoassays and genetic analysis.

3.1 CLASSICAL MICROBIOLOGY

A bacterium or virus possesses its own growing and phenotypic characteristics, and its own biochemical parameters. Characteristics such as mobility will show immediately that a bacillus cannot be, for example, *Bacillus anthracis*. Many other parameters are important. All these analyses unfortunately take time, and will require two or three days. One of the advantages is that the strain is isolated and that an antibiogram can be obtained, allowing the precise treatment of affected people. These techniques are not easy to use in the field; they require a laboratory located in advanced hospitals near the battlefield. Nevertheless, they give information which can confirm results obtained with other techniques. Things are changing in classical microbiology; on the medical diagnostic side new equipment has been developed which is monitored by computer, even if it will not be fieldable soon. It gives information on all biochemical parameters. It suffices to introduce the strain isolated into the equipment, which will analyze it and give the results with an estimation of identification (e.g. *Bacillus anthracis* identification, probability 95%). The time needed to obtain data is reduced.

One of the main advantages of these techniques is that they provide identification on the basis of phenotypic and biochemical parameters. They isolate the agent and allow other analysis like the antibiogram which can be useful to treat contaminated people. Their limitations are related to time; they are lengthy techniques and need some days in cases of bacteria and even weeks in cases of viruses that are difficult to grow. Some bacteria are very difficult to grow and these techniques can be difficult to use in very specific cases. They cannot distinguish between natural and modified microorganisms and cannot detect toxins. They need containment corresponding to their risk classification and as such can present a risk for laboratory workers.

3.2 IMMUNOASSAY TECHNIQUES

These techniques are based on the specificity of antibodies for antigens. They are very rapid and the new development of hand-held test kits (based on Enzyme Linked Immuno Sorbent Assays - ELISA) allows results to be obtained in less than 15 minutes with very high sensitivity. These techniques will confirm with high confidence the presence or not of an agent. They work perfectly with high specificity on every type of agent - toxins, viruses, bacteria - but are unable to distinguish between live and dead bacteria. These techniques are fieldable and the kits will be in use in many NATO armies; their cost will be very low, approximately five dollars each. These techniques are very specific as it is necessary to search for a precise agent but they will not identify an agent for which no antibody has been produced.

Two of the main advantages of these techniques are their rapidity and cost. They can detect every type of agent from bacteria to toxins. Their specificity can be a limitation: antibodies are directed against phenotypic molecules, fragments of membrane, proteins and toxins. It is always possible that a genetic modification might change this (e.g. in the case of the transfer of a virulent gene of one bacterium into another type of bacterium, if the antibody is not directed against the *expression* of this gene, it will not detect the new virulent bacterium). The other limitation due to specificity is that some natural bacteria can react with antibodies and give a false positive. It is necessary, in order to avoid this problem, to possess highly specific antibodies which are difficult to obtain and very expensive if developed on a small scale.

3.3 GENETIC TECHNIQUES

These techniques are as specific as immunoassay techniques. Different techniques can be used, from very simple to very sophisticated. The simplest is the Polymerase Chain Reaction (PCR); its use is very basic and permits identification of the gene coding for a toxin, or an agent by its genome. It is widely used for diagnosis of diseases, especially because PCR reagents are available in kits and it can be done in a few hours. In its scope for detection of agents in the field, it is considered as the confirmation technique when combined with the results of the other techniques. It is very sensitive and able to detect and identify from one to 10 microorganisms in a sample. These techniques will not be further elaborated on here as they are presented in other contributions to this volume.

The main advantages of genetic techniques are their high specificity and sensitivity. Their limitation is the need to determine precisely the specific target sequences of the agent that are unique. This is not very easy and is expensive. A probe can be considered specific as long as it is able to identify only the agent itself and not closely-related species. The day another bacterium is found to be detectable by the probe, it is no longer specific. PCR on pure culture is very easy, but in environmental samples it may be unable to work due to the presence of inhibitors. Special techniques to identify this problem and bypass it must be used. It has often been said that PCR cannot detect a toxin, which is right, but one must bear in mind that it can detect the organisms able to produce the toxin: to look for botulinum toxin, it is possible to use PCR and probes for *Clostridium botulinum*. These genetic techniques are not easy to use in the field; they require, for the moment, a laboratory which can be located in advanced hospitals near the battlefield. Progress will be made in that area and new detectors based on these techniques should be developed for field use in the near future.

4. Integration of Techniques to Meet the Need

4.1 ALERT DETECTION AND PHYSICAL TECHNIQUES

This section will be very short since the techniques described do not use biotechnology, but it seems necessary for a better understanding of the strategy of detection on the battlefield. At the current level of technology, alarm detection uses mainly physical means. It is impossible to use techniques to identify the agents; only the abnormal presence of live agents in the air can be detected.

The most common use of this kind of detection is to continuously analyze the air. Two solutions are possible: remote or local detection.

4.1.1 Remote detection

This may use a Lidar system, able to detect at a distance of kilometers the presence of abnormal agents in clouds. Depending on climatic conditions, this will allow some few minutes to collect air and analyze it (as described below). Lidars have been extensively investigated for chemical detection and to some extent for biological detection.

4.1.2 Local (point) detection

The second type of strategy is to use directly either particle analysers like APS or the more developed FLAPS, which analyze the number and size of particles and allow continuous monitoring of what the air contains, or to use biocollectors, able to concentrate quickly large volumes of air and at the same time able to function at a flow rate small enough to avoid destroying agents by impaction. After collection different types of techniques can be used:
- Particle detection, dependent on their size and viability criteria. Flow cytometry seems to be very promising and in some conditions can differentiate some agents such as *B. anthracis* and *Y. pestis* by their spectra.
- GC/MS (gas chromatography/mass spectrometry): A new technique coupling GC to MS seems to be promising and allows detection of bacteria able to sporulate.
- IMS (Ion Mobility System) combined with pyrolysis and gas chromatography is in favor in some countries at the moment.
- Flame photometry: This technique uses the destruction of biological agents by pyrolysis and analyzes the concentration of various ions.

4.2 CONTROL DETECTION AND IDENTIFICATION

These techniques are applied to selected agents. They can detect or identify only the agents for which specific reagents are available. They are applied at the moment to point detection of agents. This point detection could in future be delocalized using automatons able to move on the battlefield and continuously analyze the air. This will certainly be one of the main challenges in detection, allowing for warning and the taking of protection measures before the cloud reaches personnel themselves.

In control detection it is necessary to differentiate between short- and medium-term studies which will allow the development of equipment for the near future (for which the technologies are based on antigens and antibodies) and long-term studies which will allow the use in the future (10 years hence) of more sophisticated techniques such as PCR combined with physical techniques.

4.2.1 Available techniques for use on the battlefield

We will use as an example the concept of a mobile laboratory. This laboratory will be equipped with particle detectors as previously described for monitoring the surrounding air. It will include biocollectors functioning continuously and collecting air particles. The particles will be collected in liquid, generally buffer solution, which can be directly connected to detection equipment.

Hand-Held Test Kits (HHTKs). This is a very simple device produced under NATO-PG 32; it resembles a pregnancy test kit and gives an answer in less than 15 minutes. Basically, it permits the rapid monitoring and specific identification of toxins and infectious biological warfare agents at sensitivities comparable to standard ELISA format. HHTKs are defined as fast and simple-to-use, antibody-based, hand-held assays, providing unambiguous yes/no read-outs and not requiring trained operators either for sample application or for results/read-out interpretation. HHTKs will serve as a screening device, backed up by more specific, precise, and possibly time-consuming detection and identification methods. The HHTK is envisioned for use in environmental (and clinical laboratory) applications, under normal climatic and currently occurring meteorological conditions.

The current format is an immunochromatographic assay encompassing a membrane-linked, antibody-based, sandwich assay utilizing capture antibody and detector antibody reagents. The capture antibody reagent is printed as a line across a strip of nitro-cellulose membrane. The membrane has a specific pore size and the ability to bind the captured antibody within its matrix. Liquid samples migrate through and on the nitro-cellulose membrane strip at a specific rate based upon the pore size of the membrane. This assay system utilizes detector antibody coated on colloidal gold particles, size e.g. 40 nanometers. The detector antibody-labeled colloidal gold particles are deep red in color. They provide the reporter signal in the assay without the need for labile enzyme substrate reagents. Liquid sample is added to the sample well. This action places the solution in contact with the sample delivery pad. The liquid sample reconstitutes the lyophilized reagents and is delivered to the membrane. The solution wicks along the nitro-cellulose membrane. If antigen is present in the sample, it will form a complex with the detector antibody-coated colloidal gold particles, producing a red line in the test window. This assay format incorporates a reaction control into the chromatographic assay strips by printing a line of antigen specific to the species of antibody coated on the colloidal gold particles. A positive assay will have two red lines, one at the site where the capture antibody reacts with the sample antigen and one at the control line. A negative assay will have a single red line at the control site.

ELISA Test Kits. Specific devices are produced for the detection of single agents, one device per agent. They are easy to produce and somewhat complementary to other means, and can be considered as devices requiring specialized expertise due to their high sensitivity. They allow a second confirmatory check of identified agents. The first of these devices was developed during the Gulf war, for botulinum toxin. They are ELISA-based and need to be operated by specialists; any known agent can be detected by these kits.

Automatons for detection. Many types of equipment are available similar to the equipment developed under NATO-PG 33. The French ADIBio is mobile and can function in the field; power is supplied by batteries. The principle of the technique is based on ELISA, but slightly modified because tangential filtration is used to avoid clogging (ELIFA). Samples can be taken directly after biocollection or taken from the environment. They can be liquid or gas; air is collected by impact on liquid. Detection can be focused on one or more agents; cassettes allow simultaneous identification of at least five agents.

Flow cytometry. Many varieties of equipment are available which use flow cytometry. It is a combination of physical analyses (counting particles and analyzing their size and shape) and biological analyses. Repartition spectra allow differentiation between agents and the use of beads coated with antibodies also allows the identification of toxins.

4.2.2 Future techniques for use on the battlefield (some are available currently for detection in laboratories)

MiniVidas. This small device is undergoing evaluation in the laboratory. It is a very sophisticated system, ELISA-based, with a high degree of automaticity and the ability to analyze 12 samples simultaneously in one hour. Its sensitivity for the detection of SEB is 0.4 ng/ml. It could be integrated into the chain of the detection system.

PCR (Polymerase Chain Reaction). PCR is a very interesting technique which should allow identification of genes of agents when some problems have been overcome. Whilst this technique is easy to use on pure strains and solutions, environmental samples can be very complex and interfere with DNA extraction. This technique is very promising and will improve rapidly in sensitivity when combined with new techniques in addition to electrophoresis. The PCR technique is the link between the short-term studies and the long-term ones for which evaluation of new technologies is necessary.

Biosensors. Many techniques using biosensors are under development. Some seemed very promising some years ago but are still under development today. Some difficulties need to be resolved but biosensors are still promising:

- SPR (Surface Plasmon Resonance): this technique is employed by pharmaceutical companies and is very interesting for analysis and definition of antibody isotypes. It can be coupled with a sol-gel wave guide to improve performance. Very sensitive to any movement, this technique will not be used in the field for years.
- QCM (Quartz Crystal Microbalance): this technique uses a reversible piezoelectric effect. The shape of the crystal is modified by an electric field and produces an acoustic vibration whose frequency is proportional to the mass of the crystal. The frequency variation due to the binding of antigens to the antibodies previously bound to the crystal can be detected and registered.
- Capillary electrophoresis: this technique is under evaluation in detection laboratories for the rapid detection of PCR products.

5. Conclusion

To conclude this brief presentation of the current state-of-the-art of technology in the field of detection of biological agents, it must be stressed that very rapid progress is being made every day and that new techniques will be available in the near future. These techniques will certainly change the strategy set for detection. However, even though we can go far in rapidity of detection and sensitivity and complementarity of the techniques to obtain orthogonal confirmation of the nature of agents, the main challenge will remain: the achievement of early and remote detection of agents far enough from the battlefield to protect soldiers. The current automates which delocalize the point detection are a kind of 'ersatz' solution even if their production represents an important advance.

DETECTING BIOLOGICAL TERRORISM

Evaluating the Technologies

GARY EIFRIED
EAI Corporation
1308 Continental Dr
Abingdon, MD 21009

1. Introduction

As a result of the bombing of the World Trade Center in 1993, the Murrah Federal Building in 1995, and the Sarin attack on the Tokyo subway system, the United States Government initiated an extensive program to respond to the threat of terrorism in all its forms, including biological terrorism. One element of the program has been the training of emergency responders, such as police, firefighters, medical personnel and hazardous materials technicians, to recognize, respond to and recover from terrorism involving the use of weapons of mass destruction.

Our company has been involved in this effort since 1995 both privately and through the federal government program. To date, we have trained over 20,000 emergency response personnel under the U.S. Government-sponsored Domestic Preparedness Program.

This chapter summarizes some things we have learned during the course of this program. I highlight the capabilities which emergency personnel and planners themselves believe they need to respond to such incidents.

2. The Threat is Different

The character of the threat posed by bio weapons on the battlefield and bio weapons in terrorism differs. The terrorist is not limited by the same constraints regarding, for example, agent behavior. He does not have to be so concerned with specific characteristics sought in military agents such as stability in storage and non-transmission from person to person. Thus, the terrorist has a much broader range of pathogens, up to 300, from which to choose, and may simply

A. Kelle et al. (eds.), The Role of Biotechnology in Countering BTW Agents, 129–140.
© 2001 *Kluwer Academic Publishers. Printed in the Netherlands.*

select the one most easy to obtain. Table 1 is a list of potential biological agents extracted from a list prepared by Malcolm Dando [1], and supplemented by other sources [2, 3].

TABLE 1. Potential Biological Terrorism Agents [1, 2, 3]

Bacteria	Bacillus anthracis
	Brucella abortus
	Brucella meltensis
	Brucella suis
	Chlamydia psittaci
	Clostridium botulinum
	Clostridium perfringens
	Clostridium tetani
	Cornobacterium
	   Diphtheriae
		Enterohemorrhagic
		   Escherichia coli, sero-
		   type 0157 and other
		   verotoxin producing
		   serotypes
		Francisella tularensis
		Legionella pneumophila
		Malleomyces mallei
		Malleomyces pseudo-
		   mallei
		Mycobaeterium tuber-
		   culosis
			Pasteurella Tularonsis
			Pasteurella pestis
			Pseudomonas mallei
			Pseudomonas pseudo-
			   mallei
			Salmonella typhi
			Shigella dysenteriae
			Vibrio cholerae
			Yersinia pestis
			Yersinia pseudotuber-
			   culosis

Toxins	Abrin
	Botuilnum toxins
	Cholera toxin
	Clostridium perfringens
	   toxins
	Conotoxin
		Microcystins (Cyan-
		   ginosins)
		Ricin
		Saxitoxin
		Shigatoxin
		Staphylococcus aureus
		   toxin
			Tetanus toxin
			Tetrodotoxin
			Trichothecene myco-
			   toxin
			Verotoxin

Rickettsiae	Coxiella burnetil Rickettsia prowasecki	Rickettsia rickettsia Rickettsia mooseri	Rickettsia tsutugarnusni Rochalimaea quintana

Viruses	Chlkungunya virus
	Congo-Crimean
	hemorrhagic fever virus
	Dengue fever virus
	Eastern equine encepha-
	litis virus
	Ebola virus
	Hantaan virus
	Japanese encephalitis
	   virus
	Junin virus
	Kyasanur Forest virus
	Lassa fever virus
		Louping III virus
		Lymphocytic chorto-
		meningitis virus
		Machupo virus
		Marburg virus
		Monkey pox virus
		Murray Valley encepha-
		   litis virus
		Omsk hemorrhagic
		   fever viurs
		Oropouche virus
		Psittacosis
		Powassan virus
		Ritt Valley fever virus
			Rocio virus
			St. Louis encephalitis
			   virus
			Tick-borne encephalitis
			   virus (Russian Spring-
			   Summer encephalitis
			   virus)
			Varlola virus
			Venezuelan equine
			   encephalitis virus
			Western equine
			   encephalitis virus
			White pox
			Yellow fever virus

Fungi	Coccidioidie immits	Histoplasm capsulatum	Norcardia asterosides
Geneti- cally Modified Micro- organisms	Genetically modified micro- organisms or genetic elements that contain nucleic acid sequences as- sociated with pathogenicity and are derived from organisms in the core list.		Genetically modified micro- organisms or genetic elements that contain nucleic acid sequences coding for any of the toxins in the core list.

Another potential difference in biological terrorism is that of quantity. Although small quantities of agent are required in either case, there is a greater potential for use of small, impure quantities (1-2 kilogram) in the terrorist scenario.

A third set of differences includes the general organization, warning and reporting system, protective equipment and state of readiness of the civil population as contrasted to a military force.

Other differences include dissemination efficiency and purity.

Each of these areas of difference impacts what is required of a bio detection system in the terrorist situation.

3. Challenges in Biological Detection

Some of the challenges inherent in detecting biological terrorism are explained below. Many of these problems parallel those with battlefield detection.

Background count. First, there is a lot of biological material floating around in the air that we have to sort out from the biological agents of concern. This requires that our systems have the ability to discriminate, or be selective, about what they detect.

2-7 days incubation time. The incubation time is the root of the problem. It is one good reason we need detectors, since the delay between exposure and observable symptoms makes recognition of an attack very difficult. On the other hand, if we recognize an attack immediately, incubation time may give us some time to react.

Best treatment requires early identification. However, since optimal treatment requires early recognition and identification, time is still of the essence. In the case of inhalation Anthrax, for example, once major symptoms develop, treatment with antibiotics does little to alter the course of the disease.

Differentiation (natural vs. deliberate). The problem of differentiating between a naturally occurring epidemic and one caused by deliberate dissemination is difficult both on the battlefield and in the civil sector.

Training and maintaining skills. This is even more difficult in the civil sector than it is in the military. No municipalities that I have been to are able to afford full-time biological detection specialists. Any equipment will have to be simple to learn, simple to operate and simple to maintain. Some of it will need to be used in full individual protective equipment. Built-in instructions, tutorials and non-hazardous training simulants are suggested.

The final problem is time. Not only do we not have a lot of time to respond following the attack, we also don't have a lot of time to develop an approach to the problems of biological terrorism. There have been too many cases of individuals and groups seeking and obtaining biological agents under dubious circumstances not to draw the conclusion that we will likely see a major biological terrorism incident within the next few years.

4. Biological Detection Functions

If we look at biological detection in a terrorism situation in a bit more detail, we can list four specific functions over time. Those are: warn, confirm/deny, identify, and clear.

The warning function indicates that a potential biological attack may be taking place, or rather that a detection of a sudden increase in biological material has occurred. It requires a widely distributed network of inexpensive, automated sensors. The confirm or deny function determines whether or not the increase is of potentially harmful material. That is the function differentiates between a harmless increase in biological material, such as pollen, as opposed to potentially harmful pathogens. This requires fewer, but still inexpensive systems, automated in action but possibly activated on demand. The identify function identifies the organism by serotype. The clear function is used to verify that there is no longer a biological hazard. We have an existing model for part of this system, the Biological Integrated Detection System (BIDS), which addresses the warn and confirm/deny function, and to a degree, the identify function. However, while the BIDS is an important step in the right direction, only a few organisms can currently be identified, its sensitivity is not sufficient to give an "all clear" at the desired levels, it is very expensive ($1.25 M each), and it requires full-time, highly-trained operators.

5. Acceptable Performance

Figure 1 is a matrix of suggested "acceptable" performance characteristics of a bio detector for each of the detection functions. Acceptable in this case does not mean "minimum", but "necessary and adequate," or in some cases, "optimal." For example, *warning* must take place very rapidly, probably within seconds, while the process of determining an "all clear" could take much longer, with only marginal detriment to the overall attack recovery process. On the other hand, you want to be able to *confirm* an attack very quickly after warning (within minutes), but *identification* could take longer (hours).

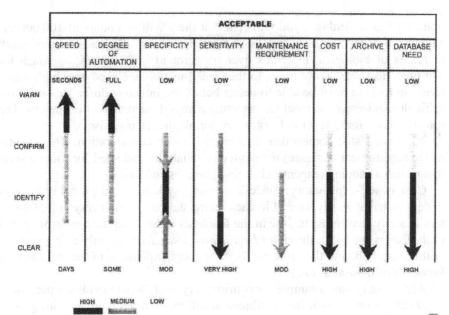

Darker intensity of the line indicates a higher level of performance is necessary

Figure 1. "Acceptable" Performance Criteria

The necessary degree of automation is high for the *warn* function, in which unattended automatic systems are desired, while increasing degrees of a man-in-the-loop are acceptable as we go through the functions *confirm, identify* and *clear.*

The degree of specificity determines how well the system can discriminate among biological materials. The *warn* function needs a system with low specificity, something generic, such as an airborne particle counter (as in BIDS), or a detector which can sense a significant increase in a common biological component. This will provide a broad trigger for warning, while not restricting that function to known threat organisms. As the function moves through *confirm* to *identify*, specificity must get very finite, but for the *clear* function, it can be less specific, since we will have already determined what the organism is.

A high degree of sensitivity is not important for *warning*. In fact, it is not desirable, since a less sensitive warning device will also be less prone to false alarms. Consider that in a typical metro station there is about 60,000 m^3 volume. A container of 355 g. of a biological agent (typical soft-drink can) will hold in the vicinity of 3.5×10^{13} organisms. Even if this is disseminated at only 0.1% efficiency, there will still be a half-million organisms per m^3 – plenty of organisms to detect. The *warn* function does not require a system which can detect just a few particles. If you calculate the concentration for an outside dissemination of 1 kg. of Anthrax, you will find that there will be counts of 100,000 organisms/m^3 over areas several kilometers downwind. Trying to detect small increases of biological materials over background is the wrong approach for warning. However, as we move to the *clear* function, we need very high sensitivity; in fact we must be able to detect below the infective dose. This is a very difficult requirement to meet for detecting many of the potential pathogens. The infective dose for Q fever is 1 organism; for plague, 10 organisms.

We have listed some other characteristics for consideration. The maintenance requirement addresses the degree of reliability and need for routine servicing, such as adding reagents, changing batteries and so on.

Cost is self-explanatory. Since a *warning* system would probably be very widely distributed, say several in each metro station, or one every 300-500 meters in a city, cost must be low in the hundreds rather than thousands of dollars each. But as we move through *identification and clearing*, only a few systems will be required, and therefore each can be more expensive, in the thousands to hundreds of thousands range.

Archiving retains a sample for confirmatory analysis and evidence purposes.

Database needs indicate the library of information required, as with genetic methods, or the diversity of reagent material needed, as with immunological methods.

You may differ with some of the specific performance levels indicated, or even want to add functions, but the overall method can be adapted to whatever criteria you select.

6.   How the Technologies Meet the Requirements

In reviewing the literature, I found reference to over 50 different approaches to biological detection [4, 5]. With the help of a paper presented by Dr. Ron Atlas at a NATO Advanced Research Workshop [6] in Budapest in 1996, I selected a variety of generic technologies and applied their performance characteristics to the functional requirements. These technologies range from the tried and true tools of microbiology to the cutting edge technology of gene probe and GC/MS. The last one is a recent request for proposal from the Defense Advanced Research Projects Agency (DARPA). I added it to see how close a system meeting its requirements comes to solving the problem.

I want to stress that I am by no means an expert in any of these technologies. I relied on the information contained in technical reviews and other literature to evaluate the degree to which each generic technology would meet the requirements. I encourage those of you who are experts to feel free to match your favorite technology to the requirements in Figure 1.

6.1 CULTURE TECHNIQUES

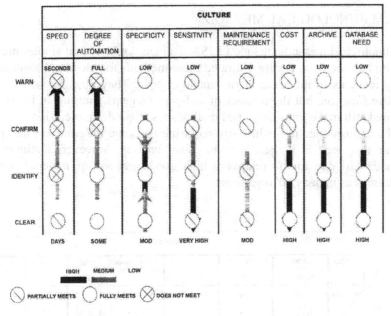

Figure 2. Performance Levels for Culture Techniques

Figure 2 shows the results for *culture* techniques. These include the classical microbiological methods used to grow colonies or quantities of viable organisms, followed by morphological, physiological and biochemical examination. Viable organisms are required, and results can be compared to a large body of epidemiological data. An advantage of these methods is that cultures are available for further analysis [6]. The woven pattern circles on the chart indicate that the technology meets the requirement at an acceptable level; slashed circles indicate a partial meeting of the requirement; and brick pattern circles indicate not meeting the requirement at an acceptable level. As the figure shows, the method has problems for all functions.

For the *warn* function, it fails because of the time and human intervention required to culture the organism, and is deficient in cost and maintenance due to the need for control of incubation conditions and the wide range of culture materials necessary. Sensitivity would not be a problem (only one viable organism is required) except for the fact that some pathogens are difficult to culture, and thus may not be detected. Similar rationale applies to the *confirm* function. The *identify* function is not met primarily because the time it takes to identify an organism by this method may exceed the incubation time in humans. You can see that while there are some deficiencies, the culture method can do reasonably well for the *clear* function.

6.2 IMMUNOLOGICAL METHODS

Immunological methods include ELISA, antigen-antibody and similar methods. As you can see from Figure 3, I rated these methods too slow and requiring too many reagents to meet the *warn* function needs. They do a bit better for the *confirm* function, but the number of different reagents and human intervention required still make immunological methods less than desirable. They fare better for the *identify* function; while still requiring the same reagents, the acceptability for this need is higher. For the *clear* function, however, immunological methods fail due to the relatively high concentrations of material required, sometimes as high as 10^7 organisms.

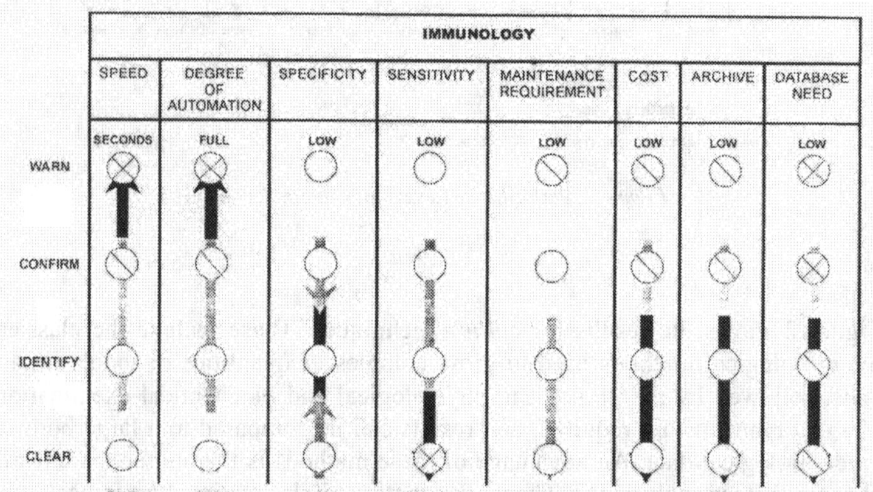

Figure 3. Performance Levels for Immunological Techniques

6.3 GENE PROBE

Figure 4 shows the analysis for *Gene Probe* methods. These methods use labeled DNA sequences of a known organism to hybridize with DNA from the sample. These methods have problems with *degree of automation, maintenance* and *cost* for the *warn* function. Similar kinds of problems exist for the *confirm* function, although the requirements are less stringent, and the ratings move to partially met. These methods do work very well for the *identify* function, although there will be a large database requirement which will be difficult to fully achieve.

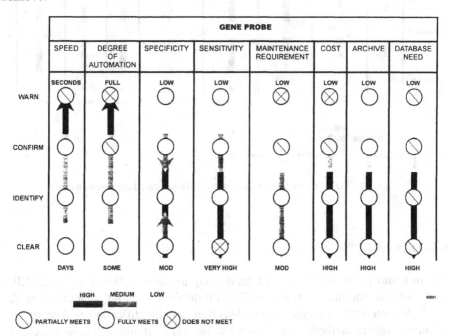

Figure 4. Performance Levels for Gene Probe Techniques

As with immunological methods, the need for very high sensitivity makes gene probe methods unsuitable for the *clear* function, since they generally require 10^6 organisms.

6.4 CHEMICAL BIOLOGICAL MASS SPECTROMETER (CBMS)

Figure 5 shows the results for the CBMS. In this method, cells are lysed and the resulting chemical component signatures determined by the instrument and compared to a spectral library. Again, this method does not seem to be appropriate for the *warn or confirm* functions, and has difficulties for the *identify* and *clear* functions.

138

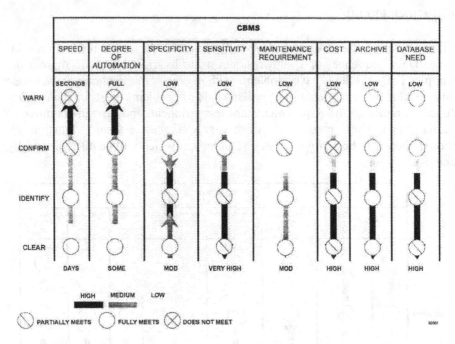

Figure 5. Performance Levels for Chemical Biological Mass Spectrometer

6.5 DARPA RFP

Figure 6 shows the results if we take the requirements outlined in the DARPA RFP and imagine that all the capabilities it describes can be achieved. The desired characteristics include "rapid detection and identification of known and unknown agents methods that do not use liquids or minimal liquids hand-held automated, dispersed, remotely placed and networked high sensitivity low false alarm with minimal sample pretreatment." IF it all works as desired, and that's a big if, we will have a system which, except for the warning function, probably meets most of the needs for a bio detection system for terrorism response. In any case, the evaluation method described in this paper will be useful in prioritizing the technologies which result from the development program.

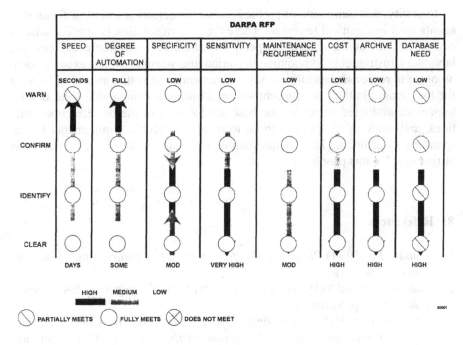

Figure 6. Performance Levels for Device Sought by DARPA

7. Conclusions

Using the method and parameters described, we can summarize the results in Table 2. We see that no single current technology evaluated meets the requirements for the *warn, confirm, identify* and *clear* functions at a reasonably satisfactory level. Worse yet, no single technology fully satisfies the requirements for any single function.

TABLE 2. Comparison of Performance Levels for All Techniques

	Culture	Immuno-logical	Gene Probe	CBMS	DARPA RFP
Warn	Red	Red	Red	Red	Amber
Confirm	Red	Red	Amber	Red	Amber
Identify	Red	Amber	Amber	Amber	Amber
Clear	Amber	Red	Red	Amber	Amber

In reality, this same kind of problem exists for detection and identification of agents used in chemical terrorism. There the current solution is to use a suite of equipment, including detector papers, wet-chemistry, electronic devices and laboratory instruments to optimize detection and warning within existing technology and response capabilities. My recommendation is that instead of looking for the "magic bullet" we concentrate on adapting a comparable suite of biological technologies, selecting one best one for each of the bio detection functions, and working to optimize these to meet the specific requirements for responding to terrorism. The key question is, what can we do within the next year or two to solve this problem?

8. References

1. Dando, M. (1994) *Biological Warfare in the 21ˢᵗ Century,* Brassey's, London & New York.
2. Gander, T.J. (ed.) (1997) *Jane's NBC Protection Equipment,* Jane's Information Group, Surrey, UK.
3. Crowell, R. (1997) *Proceedings Report, Conference on Countering Biological Terrorism: Strategic Firepower in the Hands of Many?* Potomac Institute for Policy Studies, Arlington, Virginia.
4. U.S. Army Chemical Research, Development and Engineering Center (1991) *Biological Agent Detection Systems, Final Report*, Aberdeen Proving Ground, MD.
5. U.S. Army Chemical and Biological Defense Command (1995) *Market Survey report of Biological Integrated Detection System Technologies and Equipment,* Aberdeen Proving Ground, MD.
6. Atlas, Ronald (1996) Paper presented at the NATO Advanced Research Workshop, The Technology of Biological Arms Control and Disarmament", 28-30 March, Budapest, Hungary.

DETECTING BIOLOGICAL AND TOXIN WEAPON AGENTS IN AN INSPECTION ENVIRONMENT

GRAHAM S. PEARSON
Visiting Professor of International Security
Department of Peace Studies
University of Bradford
Bradford, West Yorkshire BD7 1DP
United Kingdom

1.   Introduction

The detection of biological and toxin weapon agents in an inspection environment is a subject which is extensively debated when consideration is being given to the elaboration of the Protocol regime to strengthen the effectiveness and improve the implementation of the Biological and Toxin Weapons Convention (BTWC) [1]. This chapter considers the various inspection scenarios being considered in the Protocol to the BTWC which is nearing completion. The role that sampling and analysis – with detection and identification of biological and toxin weapon agents – can contribute to an effective regime is examined in the light of experience gained in other regimes, ranging from the UN Secretary-General investigations of alleged use to the UN Special Commission (UNSCOM) on Iraq and the Chemical Weapons Convention (CWC). It is concluded that sampling and analysis using validated analytical techniques in internationally accredited laboratories is an essential tool in an inspection environment, even though it may be used only infrequently.

2.   The BTWC Protocol

The Protocol to strengthen the effectiveness and improve the implementation of the BTWC is nearing completion. The key elements of the Protocol have all been elaborated and are now largely agreed, although there are still a few issues that have yet to be resolved. The particular elements of relevance to the Protocol inspection environment are the following:

141

A. Kelle et al. (eds.), The Role of Biotechnology in Countering BTW Agents, 141–159.
© *2001 Kluwer Academic Publishers. Printed in the Netherlands.*

a. Mandatory declarations of those activities/facilities of greatest relevance to the Convention.

b. Declaration follow-up procedures including infrequent randomly-selected visits to declared facilities to ensure consistency of declarations, declaration clarification procedures to address any ambiguity, uncertainty, anomaly or omission in a declaration which may result in a clarification visit, and assistance visits.

c. Investigations of concerns about possible non-compliance under Article I of the BTWC. Such investigations may be either field investigations or facility investigations.

d. An international organization to implement the provisions of the Protocol.

The inspection scenarios under the Protocol regime can usefully be considered in two broad groups, visits and investigations, each of which is considered in turn.

3. Visits

In the Protocol regime, visits are part of the routine provisions intended to increase transparency and enhance confidence between States Parties. Their purpose is thus primarily to ensure consistency of declarations as well as to ensure that the Protocol organization acquires and maintains a comprehensive and up-to-date understanding of the different types of facilities and activities declared globally.

There are three categories of visits:

a. Randomly-selected visits. These will be infrequent visits to declared facilities selected on a random basis. Their primary purpose is to confirm that the declarations are consistent with the declaration obligations in the Protocol.

b. Declaration clarification procedures. Provision is made for any ambiguity, uncertainty, anomaly or omission in a declaration to be resolved through a procedure in which resolution is first sought through correspondence, then through a consultative meeting on the territory of the State Party concerned, and, should the matter then still be unresolved, through a clarification visit to the facility concerned.

c. Assistance visits. Such visits will be made at the invitation of the State Party concerned in order to provide technical assistance and information or assistance in implementing the Protocol.

As might be expected, sampling and analysis is not relevant to these assistance visits although assistance may be sought to establish a capability within a State Party to carry out validated analytical techniques or to gain international accreditation for a laboratory within a State Party. Thus the August 2000 version [2] of the Protocol includes provision for the Technical Secretariat to:

> "(j) Implement at the request of States Parties, programs of support and assistance for upgrading laboratories nominated for designation and certification pursuant to Annex D, section I, part B;
> (k) Implement programs of support and assistance for designation and certification of laboratories pursuant to Annex D, section I, part B."

Insofar as randomly-selected visits are concerned, although sampling and analysis was included as a possible provision which might be offered by the visited State Party in drafts of the Protocol as late as February 2000, the current draft Protocol contains no mention of sampling and analysis in the context of randomly-selected visits. There is currently (August 2000) provision, within square brackets and thus not yet agreed, allowing the visited State Party to offer the visiting team any other on-site activities that the State Party believes may help the visiting team fulfil its mandate. It is evident that sampling and analysis has no role to play in a randomly-selected visit.

For clarification visits, the current text (August 2000) includes, also within square brackets and not yet agreed, the following language on sampling and analysis:

> "[(f) Sampling shall not be conducted unless offered by the visited State Party and visited facility personnel and deemed useful by the visiting team. Any mutually agreed sampling and analysis shall be performed by facility personnel in the presence of the visiting team and representatives of the visited State Party. The visiting team shall not seek to remove samples from the facility.]"

In addition, also in square brackets, is comparable language to that in regard to randomly-selected visits stating that:

> "[(e) The visited State Party may [, at its own initiative or at the suggestion of the visiting team,] offer the visiting team, at any time during the visit, any other on-site activities which the visited State Party believes may assist the visiting team to fulfil its mandate;]"

The option should be available, should the visited State Party wish, to offer sampling and analysis as this may in the circumstances of some clarification visits provide the easiest and fastest way of resolving an uncertainty.

4. Investigations

In the Protocol regime, each State Party has the right to request an investigation to determine the facts relating to a specific concern about possible non-compliance with the Convention by any other State Party. There are two principal types of investigation:

a. Field investigation. Investigations to be conducted in geographic areas where exposure of humans, animals or plants to microbial or other biological agents and/or toxins has given rise to a concern about possible non-compliance under Article I of the Convention or alleged use of biological weapons;

b. Facility investigation. Investigations of alleged breaches of obligations under Article I of the Convention, to be conducted inside the perimeter of a particular facility at which there is a substantiated concern that it is involved in activities prohibited by Article I of the Convention.

In the case of field investigations, the Protocol includes provisions for both disease/intoxination-related examinations and for sampling and identification, whereas for facility investigations there are provisions for examination of clinical and pathological samples (which are currently within square brackets) and for sampling and identification. The provisions for sampling and identification in the two types of investigation are broadly similar but differ in detail.

4.1 FIELD INVESTIGATIONS

The provisions (August 2000) for disease/intoxination-related examination relating to sampling state that:

> "35. The investigation team may, where necessary and applicable, take body samples from affected persons or animals as well as samples of affected or exposed plants in order to diagnose, confirm a clinical diagnosis of the disease or determine whether exposure has occurred. In the case of persons affected this shall be with the informed written consent or with the informed written consent of the family or legal representative of the person affected. The receiving State Party shall receive duplicate samples for its own analysis."

In addition there is provision for the investigation team to participate in or to conduct post mortem examinations with the informed written consent of the family of the deceased, to conduct disease/intoxination-related examinations of animals and/or plants affected or exposed, with relevant explicit consent where possible and appropriate, of the legal owners of the animals and/or plants, and to examine laboratory animals or existing samples taken from laboratory animals or take samples from such animals with the consent of the legal owner. The provisions go on to state that:

> "38. All medical information, including samples and other material taken from humans, shall be accorded the most stringent protection measures by the investigation team and all laboratories involved in the investigation."

The sampling and identification provisions state (August 2000) that:

> "41. The investigation team may [with the consent of the receiving State Party], where appropriate and it considers necessary, take environ-

mental samples, samples of munitions and devices or remnants of munitions and devices relevant to the investigation mandate. Any such samples shall be analyzed for the presence of specific biological agents or toxins.

42. Samples shall be taken in the presence of a representative of the receiving State Party. The investigation team may request the receiving State Party to assist in the collection of samples under the supervision of members of the investigation team. The investigation team may also request the receiving State Party, where necessary and appropriate, to take relevant control samples from areas immediately adjacent to the locations under investigation. The receiving State Party shall receive duplicate samples for its own analysis.

43. The investigation team may analyze samples using any methods specifically designed or approved for use in such investigations, and available to the investigation team. At the request of the investigation team, the receiving State Party shall, to the extent possible, provide assistance for the analysis of samples, using locally available resources. If the receiving State Party itself performs analyses, the investigation team or some member especially assigned by the team leader shall be present during all analytical processes. All sampling shall be conducted according to procedures and methods so as to ensure that the desired samples taken are not contaminated and taken with due regard to health and safety considerations.

44. Analysis [of one of the sealed duplicate samples referred to in paragraph 42] shall, whenever possible, be carried out on the territory of the receiving State Party and in the presence of representatives of the investigation team and the receiving State Party.

45. When it is not possible to carry out the analysis on the territory of the receiving State Party, the investigation team may remove samples for analysis in designated and certified laboratories. Representatives of the receiving State Party shall have the right to accompany all samples and observe any analysis and the subsequent destruction. Any samples remaining after analyses that have not been destroyed shall be returned to the State Party of origin.

46. The Director-General shall have the primary responsibility for the security, integrity and preservation of samples and for ensuring that the confidentiality of samples transferred for off-site analysis is protected. The Director-General shall, in any case:
    (a) Establish a stringent regime governing the collection, handling, storage, transport and analysis of samples;
    (b) Select from among the designated and certified laboratories those which shall perform analytical or other functions in relation to the investigation;
    (c) Ensure that there are procedures for the safekeeping and maintaining of the integrity of sealed duplicate samples for further clarification if necessary;
    (d) Ensure the expeditious processing of the analysis of samples;
    (e) Be accountable for the safety of all samples.

47. When off-site analysis is to be performed, samples shall be analyzed in two designated and certified laboratories [in different States Parties]. The Technical Secretariat shall ensure the expeditious processing of the analysis.

48. The receiving State Party shall receive duplicate samples for its own analysis. The receiving State Party and the investigation team shall also receive sealed duplicate samples for safekeeping and use if necessary for further clarification.

49. If further clarification of analytical results becomes necessary, then the sealed duplicate samples shall be used for this purpose. The seals of these samples shall be broken in the presence of both the investigation team and representatives of the receiving State Party. The analysis of these samples shall also take place in the presence of the investigation team and representatives of the receiving State Party.

50. Any unused samples or portions thereof, remaining after the investigation has been completed and that have not been destroyed, shall be returned to the receiving State Party."

4.2 FACILITY INVESTIGATIONS

The provision, currently (August 2000) within square brackets, for the examination of clinical and pathological samples states that:

"49. The investigation team may with the permission of the receiving State Party examine analytical data related to clinical and pathological samples relevant to the investigation mandate taken previously by the facility.]"

Previous versions of the Protocol had included language allowing the investigation team to analyze such samples.

In regard to sampling and identification, the provisions, although similar to those for field investigations, are more qualified:

"49. The investigation team may [, as a last resort,] [, if required to fulfil its mandate,] request samples and test these for the presence of specific biological agents or toxins in order to address a specific non-compliance concern contained in the investigation mandate.

50. Sampling shall only be used when the investigation team comes to a conclusion [based only on information obtained from the briefing and/or the application of the other measures in this section] during the investigation which suggest that sampling might provide significant information necessary for the fulfillment of the investigation mandate. [Where possible,] specific tests shall be used to identify specific agents, strains or genes.

51. The receiving State Party shall have the right to take measures, in accordance with the access provisions contained in Article III, section G, subsection H, to protect national security and confidential proprietary information such as requiring the use of specific tests or on-site analysis or, if necessary, to refuse a sample. In the latter case

the receiving State Party shall be under the obligation to make every reasonable effort to demonstrate that the requested sample is unrelated to the non-compliance concern(s) contained in the investigation mandate.

52. Representatives of the receiving State Party shall take samples at the request of the investigation team and in their presence. If so agreed, the investigation team may take samples itself. Where possible, samples shall be analyzed on site. The investigation team may test samples using any methods approved by the Technical Secretariat for use in such investigations. At the request of the investigation team, the receiving State Party shall to the extent possible provide assistance for the analysis of samples on site, using locally available resources. In the event that it is agreed between the investigation team and the receiving State Party, that the receiving State Party itself performs analyses, this shall be done in the presence of members of the investigation team.

53. If on-site analysis is impossible, the investigation team may request the removal of samples for analysis in laboratories selected in accordance with paragraph 54 (b) below. Where possible a sample [shall] [may also] be analyzed in an accredited and certified laboratory on the territory of the receiving State Party. The receiving State Party shall have the right to take measures necessary to ensure that commercial proprietary or national security information would not be jeopardized by the off-site analysis of samples. If the removal of samples is agreed, the receiving State Party shall have the right to accompany the sample and observe any analysis and its subsequent destruction.

54. The Director-General shall have the primary responsibility for the security, integrity and preservation of samples and for ensuring that the confidentiality of samples transferred for off-site analysis is protected. The Director-General shall, in any case:
    (a) Establish a stringent regime governing the collection, handling, storage, transport and analysis of samples;
    (b) Select from among the designated and certified laboratories those which shall perform the analytical functions in relation to the investigation;
    (c) Ensure that there are procedures for the safekeeping and maintaining of the integrity of sealed duplicate samples for further clarification if necessary.

55. When off-site analysis is to be performed, samples shall be analyzed in [a] [at least two] designated and certified laborator[y][ies]. The Technical Secretariat shall ensure the expeditious processing of the analysis. The samples shall be accounted for by the Technical Secretariat.

56. The receiving State Party shall receive duplicate samples, for its own analysis. The receiving State Party and the investigation team shall also receive sealed duplicate samples for safekeeping and use if necessary for further clarification.

57. If further clarification of analytical results becomes necessary then the sealed duplicate samples shall be used for this purpose. The seals of these samples shall be broken in the presence of both the investigation team and representatives of the receiving State Party. The analysis of these samples shall also take place in the presence of the investigation team and representatives of the receiving State Party.

58. Any unused samples or portions thereof, remaining after the investigation has been completed and that have not been destroyed shall be returned to the receiving State Party.

59. The receiving State Party shall have the right to offer a sample for analysis in accordance with the provisions in paragraphs 50 to 58 above, at any time in order to help resolve the non-compliance concern(s) contained in the investigation mandate.

60. Any on-site sampling and analysis shall be conducted in such a way as to avoid any adverse impact on the normal work of the facility and any consequent loss of production."

The basic approach, in both field and facility investigations, is to require the taking of multiple samples so that duplicate analysis can be carried out in two designated and certified laboratories and to enable the receiving State Party to receive a duplicate set of samples which it can analyze if it so wishes.

4.3 DESIGNATION AND CERTIFICATION OF LABORATORIES

In respect of both types of investigations, there are provisions for the designation and certification of laboratories:

"(B) DESIGNATION AND CERTIFICATION OF LABORATORIES

18. The Director-General shall utilize only properly designated and certified laboratories for off-site analyses of samples. [Analysis [of a part of a sample] shall, whenever possible, be carried out on the territory of the receiving State Party.]

19. The criteria, including the proficiency standards, and procedures required for designation and certification of laboratories shall be approved by the First Conference of States Parties.

20. Not later than 30 days after the conclusion of the first Conference of States Parties, or after the accession of a State Party to the Protocol, the Technical Secretariat shall communicate to the States Parties the criteria, including the proficiency standards, and procedures required for the designation and certification of laboratories as approved by the First Conference of States Parties.

21. States Parties, wishing to do so, shall, within 60 days after receiving the communication of the criteria, including the proficiency standards, and procedures required for the designation and certification of laboratories, provide an initial list of laboratories nominated for designation and certification.

22. Nominated laboratories shall be designated and certified by the Director-General in accordance with the provisions contained in para-

graphs 19 and 20 above. The Director-General shall not later than 30
days after the completion of the designation and certification process, communicate a list of all the designated and certified laboratories to all States Parties.

23. The Director-General may terminate the designation and certification of a laboratory on the request of the nominating State Party or if
such a laboratory falls below the required proficiency standards.

24. Further laboratories may, when necessary, be designated and certified in accordance with the procedures referred to in paragraphs 19
to 21 above. The designation and certification of each laboratory
shall be subject to renewal every three years.

25. In the designation and certification of laboratories, the Director-
General shall pay due regard to the necessity of equitable geographic
distribution of designated laboratories. At the request of a State
Party, the Technical Secretariat shall assist in the upgrading of a
laboratory(ies) nominated for designation and certification. The cost
of upgrading the nominated laboratories shall be borne by the State
Party concerned, and/or by the Technical Secretariat within available
resources when possible.

26. In order to ensure the security and confidentiality of samples being
analyzed, the Director-General shall enter into specific agreements
with designated and certified laboratories as soon as possible after
the designation and certification of each laboratory. A designated
and certified laboratory shall not be used for the analysis of samples
until such an agreement has been concluded with the laboratory."

These provisions rightly require that a designated and certified laboratory
meets internationally agreed standards as it is very clear that the Protocol regime would rapidly fall into disrepute if samples were not to be analyzed to the
highest internationally agreed standards.

5. Other Inspection Regimes

There are three existing inspection regimes from which experience can be
drawn: the UN Secretary-General investigations of alleged use, the UN Special
Commission (UNSCOM) on Iraq, and the Chemical Weapons Convention
(CWC). Each of these is considered in turn.

5.1 UN SECRETARY-GENERAL INVESTIGATIONS OF ALLEGED USE

During the past couple of decades there have been occasions when the United
Nations Secretary-General, at the invitation of the State on whose territory the
alleged incident has occurred, has dispatched small teams of international experts to investigate the alleged use of chemical weapons. There were several
such investigations during the Iran-Iraq war of the 1980s. In general these have

been carried out using ad hoc procedures which have been used within the constraints of the particular circumstances of the incident and have depended entirely upon the support of the inviting State. The outcomes of such investigations have frequently been somewhat ambiguous because of the particular circumstances pertaining to the incident being investigated.

Allegations of the continued use of chemical weapons in the Iran-Iraq war and concern expressed by the United Nations General Assembly over the threat posed to international peace and security by the risk of the use of chemical weapons led the General Assembly on 30 November 1987 to adopt resolution 42/37 C, which

> "Requests the Secretary-General to carry out investigations in response to reports that may be brought to his attention by any Member State concerning the possible use of chemical and bacteriological (biological) or toxin weapons that may constitute a violation of the 1925 Geneva Protocol or other relevant rules of customary international law in order to ascertain the facts of the matter, and to report promptly the results of any such investigation to all Member States;"

In addition, the resolution:

> "Requests the Secretary-General, with the assistance of qualified experts provided by interested Member States, to develop further technical guidelines and procedures available to him for the timely and efficient investigation of such reports of the possible use of chemical and bacteriological (biological) or toxin weapons;"

The Secretary-General appointed a group of six qualified experts from Bulgaria, Egypt, France, the Soviet Union, Sweden and the United States which met in Geneva first from 15 to 18 August 1988 and then again from 6 to 17 February 1989 and 31 July to 11 August 1989. The report of the qualified experts to the Secretary-General was circulated [3] by the Secretary-General to the General Assembly as document A/44/561 on 4 October 1989. In this document, the Secretary-General in paragraph 7 noted that the recommendations contained in the report are those of the experts themselves and that the Secretary-General wished "to point out that, with respect to the complex and technical issues covered by the report, he is not in a position to pass judgement on all aspects of the work accomplished by the experts."

The report of the experts addresses *inter alia* the requirement for analytical laboratories where it states that:

> "76. Analytical laboratories, designated by Member States to the Secretary-General, may be called upon to carry out the following tasks: identification of CBT agents, their characteristic impurities, and degradation products, and munitions which may be related to the possible use of CBT weapons; validation of preliminary analyses; elucidation of the nature of unknown CBT agents and timely preparation and transmission of a report of the details and results of their analyses to the Secretary-General.

77. The designated laboratories may be called upon by the Secretary-general to participate in interlaboratory calibration studies so as to establish the validity and accuracy of their analytical methods in order to ensure the best expertise necessary in the samples received from the site of alleged use.

78. The laboratories may present methodologies for sample collection, transport or analysis that they have developed that may represent an improved capability, and should forward all pertinent documentation to the Secretary-General.

79. The laboratories may, if the qualified experts return to the site of the alleged incident for a follow-up investigation, ask that new types of samples be taken."

The report of the experts also provides for standing preparatory measures for investigations, which requires for the Secretary-General to carry out interlaboratory calibration studies with a view to

"first, demonstrating competence on the part of the individual designated laboratories for the detection and identification of known CBT agents; second, evaluating the capability of the individual laboratories to detect the presence of other toxic substances unknown to the laboratory in biomedical and environmental samples; and third, demonstrating the level of competence represented by the laboratories collectively for the analysis of all types of samples that may require analysis in the course of the investigation."

The detailed technical procedures for the investigation include provisions for sampling which specify that

"The qualified experts should prepare, as possible, three sets of samples collected, in two parts, as follows. One part of each set should contain samples that may be contaminated and comparable uncontaminated control samples. This part of the samples set should be identified for the laboratories as possibly contaminated. It is important that the laboratories be unable to distinguish those control samples from the other samples. The other part of each set should contain uncontaminated samples, and should be identified for the laboratory as such for the purposes of allowing the laboratory to carry out background and calibration studies for their equipment and analytical procedures;"

They go on to state that:

"After proper labeling, packing and sealing, the samples and control samples should be transported as soon as possible to three designated laboratories. Of these, two laboratories should be requested to carry out immediately the analyses required for the investigation. The third laboratory should be requested to carry out the analyses required for the investigation only if the results obtained by the two laboratories are inconclusive or contradictory, or if other circumstances exist or arise which would warrant the analysis;"

The Appendices to the report include, as Appendix V, a list of laboratory specializations which states that the designated laboratories should be capable of conducting certain specified analyses on all samples relevant to an investigation. This list includes:

"3.   Identification, in all the types of sample, of biological warfare agents (bacteria, viruses, others) and/or toxins, known and unknown.

5.   Evaluation of the effects of biological warfare agents and toxins, including epidemiological and ecological evaluation.

6.   Pathological and biochemical examination of organs and tissue taken from victims of CBT weapons, and where possible identification of the agent concerned.

7.   Pathological and biochemical examination of organs and tissue taken from animals affected by CBT weapons, and where possible identification of the agent concerned.

8.   Examination of plant tissue affected by CBT weapons, and where possible identification of the agent concerned."

It is thus evident that the framework developed by the qualified experts and reported to the Secretary-General was soundly based. However, the framework was taken no further although some States notified to the Secretary-General information on expert consultants, qualified experts and laboratories. For effective and timely investigations, the detailed provisions for the carrying out of such an investigation need to have been agreed **in advance** by the member States with the qualified experts trained and the designated laboratories established with validated analytical techniques. Consequently, there is a clear need for an international organization – such as that to be established under the Protocol to the BTWC or that established under the CWC – the OPCW – to establish the capabilities for such investigations of the use of biological, toxin or chemical weapons.

5.2   UN SPECIAL COMMISSION (UNSCOM) ON IRAQ

The UN Special Commission (UNSCOM) on Iraq was established under Security Council resolution 687(1991) [4] to carry out on-site inspection in Iraq, to supervise the destruction, removal or rendering harmless of

"(a)   All chemical and biological weapons and all stocks of agents and all related subsystems and components and all research, development, support and manufacturing facilities;"

and to provide ongoing monitoring and verification of Iraq's compliance with the Security Council decision that Iraq should not use, develop, construct or acquire chemical or biological weapons. During the years 1991 to 1998, UNSCOM succeeded, despite Iraq's steadfast refusal to cooperate, in unearthing evidence of the Iraqi chemical and biological weapons [5] In carrying out this task, UNSCOM was supported by people, equipment and facilities as well as funds provided by member States of the United Nations.

Sampling and identification of chemical and biological agents was an important tool that was used by the UNSCOM inspectors on several occasions when it considered that such analysis would provide useful information. A particular example of the value of sampling and identification of biological agents related to work carried out by UNSCOM in relation to the Iraqi accounting for the weaponization of biological weapons. Iraq in its Full, Final and Complete Declaration (FFCD) of September 1997 had declared that 25 Al Hussein type missiles were weaponized with BW agents, which Iraq had stated were filled as follows: 5 with anthrax, 16 with botulinum toxin and 4 with aflatoxin. UNSCOM had taken samples from the excavated remnants of destroyed Al Hussein warheads and the analysis of these samples in April and June 1998 revealed a serious misrepresentation in Iraq's FFCD as "traces of anthrax had been identified on remnants of at least 7 separate missile warheads". UNSCOM had subsequently, in July 1998, presented the results of its analyses to Iraq. In response, Iraq had stated that "instead of the declared 5 anthrax and 16 botulinum toxin warheads, there had been in fact 16 anthrax and 5 botulinum toxin missile warheads filled" and had said that "this change in disclosure would not affect Iraq's declaration on the total quantity of BW agents produced and weaponized".

A letter on 3 September 1998 from the Permanent Representative of Iraq to the President of the Security Council reaffirmed that the Iraqi FFCD had stated that 16 missiles had been filled with agent A, 4 with agent C and 5 with agent B and noted that UNSCOM had informed Iraq that laboratory analysis of samples of the warhead remnants showed that at least seven warheads had traces of anthrax (agent B) in them. The Iraqi response was that "when reviewing Iraq's biological weapons full, final and complete disclosure he had noticed that in view of the known quantity of agents produced in 1990 and the quantity of agents transferred to the filling station the alternative of 16 B, 4 C and 5 B (sic) could be equally valid in the absence of a definitive document".

Overall, it is evident that the role of sampling and identification had been a valuable tool needed by UNSCOM to underpin the on-site inspections [6]. However, UNSCOM did not set out and publish its procedures for sampling and identification of chemical or biological agents as it depended largely on the availability of experts and laboratories offered by member States.

5.3 CHEMICAL WEAPONS CONVENTION (CWC)

The Chemical Weapons Convention (CWC) [7] opened for signature in 1993 and entered into force on 29 April 1997. As of September 2000, 174 States have signed, ratified or acceded to the Convention and there are 139 States Parties. This Convention is of particular relevance when considering the detection of biological and toxin weapon agents in an inspection environment as the prohibition in the CWC extends to all chemicals, however produced, and thus includes toxins. This is made clear by the definition of chemical weapons in Article II which includes a general purpose criterion, emphasized in bold below, and states that:

> ""Chemical Weapons" means the following, together or separately:
> (a) Toxic chemicals and their precursors, **except where intended for purposes not prohibited under this Convention, as long as the types and quantities are consistent with such purposes;**
> (b) Munitions and devices, specifically designed to cause death or other harm through the toxic properties of those toxic chemicals specified in subparagraph (a), which would be released as a result of the employment of such munitions and devices;..."

The definition of toxic chemical, again in Article II, is all-embracing:

> ""Toxic Chemical" means:
> Any chemical which through its chemical action on life processes can cause death, temporary incapacitation or permanent harm to humans or animals. This includes all such chemicals regardless of their origin or their method of production, and regardless of whether they are produced in facilities, in munitions or elsewhere."

The CWC regime, if the provisions for destruction of chemical weapons and of chemical weapon production facilities are ignored, is broadly similar to that being developed for the Protocol to the BTWC, with the principal CWC elements being:

- a. Mandatory declarations of facilities producing Scheduled chemicals and discrete organic chemicals;
- b. Routine inspections of declared facilities;
- c. Challenge inspections of non-compliance concerns and investigations of alleged use; and
- d. An organization to implement the Convention.

5.3.1 Specific CWC provisions regarding sampling

The Verification Annex to the CWC contains the following provisions relating to samples, which in contrast to the draft BTWC Protocol are not nearly as detailed:

> *"Collection, handling and analysis of samples*
>
> 52. Representatives of the inspected State Party or of the inspected facility shall take samples at the request of the inspection team in the presence of inspectors. If so agreed in advance with the representatives of the inspected State Party or of the inspected facility, the inspection team may take samples itself.
> 53. Where possible, the analysis of samples shall be performed on-site. The inspection team shall have the right to perform on-site analysis of samples using approved equipment brought by it. At the request of the inspection team, the inspected State Party shall, in accordance with agreed procedures, provide assistance for the analysis of sam-

ples on-site. Alternatively, the inspection team may request that appropriate analysis on-site be performed in its presence.

54. The inspected State Party has the right to retain portions of all samples taken or take duplicate samples and be present when samples are analyzed on-site.

55. The inspection team shall, if it deems it necessary, transfer samples for analysis off-site at laboratories designated by the Organization.

56. The Director-General shall have the primary responsibility for the security, integrity and preservation of samples and for ensuring that the confidentiality of samples transferred for analysis off-site is protected. The Director-General shall do so in accordance with procedures, to be considered and approved by the Conference pursuant to Article VIII, paragraph 21 (i), for inclusion in the inspection manual. He shall:

   (a) Establish a stringent regime governing the collection, handling, transport and analysis of samples;

   (b) Certify the laboratories designated to perform different types of analysis;

   (c) Oversee the standardization of equipment and procedures at these designated laboratories, mobile analytical equipment and procedures, and monitor quality control and overall standards in relation to the certification of these laboratories, mobile equipment and procedures; and

   (d) Select from among the designated laboratories those which shall perform analytical or other functions in relation to specific investigations.

57. When off-site analysis is to be performed, samples shall be analyzed in at least two designated laboratories. The Technical Secretariat shall ensure the expeditious processing of the analysis. The samples shall be accounted for by the Technical Secretariat and any unused samples or portions thereof shall be returned to the Technical Secretariat.

58. The Technical Secretariat shall compile the results of the laboratory analysis of samples relevant to compliance with this Convention and include them in the final inspection report. The Technical Secretariat shall include in the report detailed information concerning the equipment and methodology employed by the designated laboratories."

Insofar as challenge inspections are concerned, the provisions in Part X of the Verification Annex make it clear that the inspection team and the inspected State Party shall negotiate the particular inspection activities, including sampling, to be carried out:

"47. The inspected State Party shall designate the perimeter entry/exit points to be used for access. The inspection team and the inspected State Party shall negotiate: the extent of access to any particular place or places within the final and requested perimeters as provided in paragraph 48; the particular inspection activities, including sampling, to be conducted by the inspection team; the performance of particular activities by the inspected State Party; and the provision of particular information by the inspected State Party."

For investigations of alleged use, the provisions in Part XI of the Verification Annex regarding sampling and analysis are clear:

"*Sampling*

16. The inspection team shall have the right to collect samples of types, and in quantities it considers necessary. If the inspection team deems it necessary, and if so requested by it, the inspected State Party shall assist in the collection of samples under the supervision of inspectors or inspection assistants. The inspected State Party shall also permit and cooperate in the collection of appropriate control samples from areas neighboring the site of the alleged use and from other areas as requested by the inspection team.

17. Samples of importance in the investigation of alleged use include toxic chemicals, munitions and devices, remnants of munitions and devices, environmental samples (air, soil, vegetation, water, snow, etc.) and biomedical samples from human or animal sources (blood, urine, excreta, tissue etc.).

18. If duplicate samples cannot be taken and the analysis is performed at off-site laboratories, any remaining sample shall, if so requested, be returned to the inspected State Party after the completion of the analysis."

The provisions relating to the report of such an investigation of alleged use make particular reference to the results of sampling and analysis:

"25. The final report shall summarize the factual findings of the inspection, particularly with regard to the alleged use cited in the request. In addition, a report of an investigation of an alleged use shall include a description of the investigation process, tracing its various stages, with special reference to:
     (a) The locations and time of sampling and on-site analyses; and
     (b) Supporting evidence, such as the records of interviews, the results of medical examinations and scientific analyses, and the documents examined by the inspection team.

26. If the inspection team collects through, inter alia, identification of any impurities or other substances during laboratory analysis of samples taken, any information in the course of its investigation that might serve to identify the origin of any chemical weapons used, that information shall be included in the report."

5.3.2 CWC Implementation Experience

In October 2000, some three and a half years after entry into force, there has been no challenge inspection or investigation of alleged use under the CWC although there have been practices of both. It is, however, clear that the Technical Secretariat has paid considerable attention to the designation of laboratories for the carrying out of off-site analyses and for demonstrating through proficiency testing, once a year, that these laboratories have maintained their capabilities. In August 2000, the Technical Secretariat advised [8] the States Parties of the CWC that 12 laboratories in 12 countries (China, Czech Republic, Finland, France, Germany, Republic of Korea, The Netherlands, Poland, Sweden, Switzerland, the UK and the US) had retained their designation status and that one laboratory (in the Russian Federation) had been newly designated. The same note reported the performance rating of the designated laboratories in the first seven official proficiency tests and noted that three of the designated laboratories had not met the criterion to receive and analyze authentic samples from the OPCW. The evaluation of the results of proficiency testing are carried out in accordance with a standard operating procedure [9] which has been developed by the OPCW in conjunction with the States Parties.

6. Conclusion

Although there has been much debate about the possible role of sampling and analysis in an inspection environment, there has generally been less consideration of the circumstances under which sampling and analysis is likely to be used as an inspection tool. There are three categories of likely scenarios in the context of the Protocol to the BTWC:

a.   Field investigations under the BTWC Protocol, which will generally be investigations of alleged use, when sampling and analysis will be necessary in order to determine the identity of the agent causing the outbreak and to identify the strain of the agent which may indicate whether or not this strain is endemic to the region;

b.   Instances during a facility investigation under the BTWC Protocol when the investigation team seeks additional information to confirm suspicions about non-compliance obtained from other inspection tools; and,

c.   Instances when a State Party being investigated under the BTWC Protocol offers sampling and analysis as a means of demonstrating its compliance.

Sampling and analysis is thus an inspection tool that will be infrequently used. It is, however, an important tool with a significant deterrent value as a State Party contemplating a breach of the Convention could not be certain that sampling and analysis might not find traces of a biological or toxin agent as was successfully demonstrated by UNSCOM in its uncovering of the Iraqi biological weapons program. In order to maximize this deterrent value, it is important not to prescribe which identification techniques may be used as the Protocol organization should be able to utilize the most advanced techniques once validated analytical procedures have been developed.

Although sampling and analysis may be an infrequently used tool, it is vital, when it is used, to ensure that there are rigorous and proven procedures for the collection of replicate samples and control samples and the chain of custody from the taking of the samples through to their analysis by validated laboratories in accredited laboratories in more than one State Party. This is necessary to ensure that the inspection regime does not fall into disrepute through the use of imprecise sampling and analysis techniques that are not accepted as valid and to the highest international standards by the States Parties.

7. References

1. United Nations General Assembly Resolution 2826 (XXVI) (1972) *Convention on the Prohibition of the Development, Production and Stockpiling of Bacteriological (Biological) and Toxin Weapons and on Their Destruction*, 16 December 1971. General Assembly, Official Records: Twenty-Sixth Session, Supplement No. 29 (A/8429), United Nations, New York.

2. United Nations (2000) *Procedural Report of the Ad Hoc Group of the States Parties to the Convention on the Prohibition of the Development, Production and Stockpiling of Bacteriological (Biological) and Toxin Weapons and on their Destruction*, BWC/AD HOC GROUP/52 (Part I), 11 August, Geneva.

3. United Nations General Assembly (1989) *Chemical and Bacteriological (Biological) Weapons, Report of the Secretary-General*, A/44/561, 4 October.

4. United Nations Security Council (1991) *Security Council Resolution establishing detailed measures for a cease fire, including deployment of a United Nations Observer Unit; arrangements for demarcating the Iraq-Kuwait borders; the removal or destruction of Iraqi weapons of mass destruction and measures to prevent their reconstitution, under the supervision of a Special Commission and Director General of the IAEA; and creation of a compensation fund to cover direct loss and damage resulting from Iraq's invasion of Kuwait*, S/RES/687(1991) 3 April.

5. Pearson, G. (1999) *The UNSCOM Saga: Chemical and Biological Weapons Non-Proliferation*, Macmillan/St. Martin's Press, New York.

6. Spertzel, R.O. (1997) Lessons from the UNSCOM experience with sampling and analysis, in Jonathan B. Tucker (ed.), *The Utility of Sampling and Analysis for Compliance Monitoring of the Biological Weapons Convention*, Report CGSR-97-001, Center for Global Security Research, February, pp. 17-25.

7. Organization for the Prohibition of Chemical Weapons, Convention on the Prohibition of the Development, Production, Stockpiling and Use of Chemical Weapons and on their Destruction. Available on the web at http://www.opcw.org.

8. Organization for the Prohibition of Chemical Weapons (2000) Note by the Director General, Status of Laboratories Designated for the Analysis of Authentic Samples, S/204/2000, 4 August.
9. Organization for the Prohibition of Chemical Weapons (1998) Note by the Director General, Revised Standing Operating Procedure for Evaluation of the Results of OPCW Proficiency Tests, S/46/98, 21 April.

POSSIBILITIES AND LIMITATIONS OF VACCINATION

KATHRYN NIXDORFF
Department of Microbiology and Genetics
Darmstadt University of Technology
Schnittspahnstr. 10
D-64287 Darmstadt
Germany

1. Introduction

The development of vaccines has been one of the most important endeavors in the fight to counter and control infectious diseases. Vaccination is based on the activation of a protective immune response to the infectious agent by immunogenic antigens of the pathogenic microorganism. Most microorganisms are not pathogenic but are indeed beneficial and can enrich our lives. Some species or strains among the microorganisms in the categories including viruses, bacteria, fungi, and protozoa can, however, produce disease. The processes by which such microorganisms cause infectious diseases are not in many cases completely understood. Generally, these organisms must be able to either colonize or invade the tissues and to cause damage. In this respect, many disease-causing microorganisms produce toxins, which are poisonous products of their metabolism.

Some of the most intensive research in the area of infectious diseases today involves the elucidation of the mechanisms of pathogenesis on the molecular biological level. In order to combat infectious diseases effectively, it is essential to be able to understand the mechanisms of the processes leading to illness as well as their regulation. An enormous amount of information has been gained in just the last decade. It becomes ever clearer that many different factors play a role in virulence [1, 2] and no one system is as yet understood in its entirety. Thus, while one factor, such as the production of a toxin, might be essential for the pathogenic process, it is in most cases only effective in concert with a variety of other, sometimes less well-defined factors that allow the microorganism to invade the host, establish itself, and multiply.

Traditionally, vaccine development has involved the attenuation or inactivation of whole organisms to produce safe and effective prophylactic reagents. Today there are strategies available that allow the precise deletion of virulence

A. Kelle et al. (eds.), *The Role of Biotechnology in Countering BTW Agents*, 161–183.
© 2001 *Kluwer Academic Publishers. Printed in the Netherlands.*

genes as well as the introduction of genetic lesions into the genome of the pathogen in order to produce a vaccine consisting of live, attenuated microorganisms [3]. These attenuated microorganisms can also act as carriers for genes of another pathogen. The strategy here is the targeting of these microorganisms to appropriate tissues in which the genes will be expressed and the products will evoke a protective immune response. Other manipulations permit the production of vaccines based on proteins or peptides (subunits) of the microorganisms, against which a protective immune response can be mounted. A survey of the literature over the past ten years reveals an enormous amount of activity in the area of vaccine development using the techniques of genetic engineering. The goal of this research is the production of vaccines in forms that are safer (cause fewer side effects) and more effective than those produced by conventional methods [4].

In order to be able to appreciate the strategies designed for vaccine development, an understanding of the immune system and how it reacts to pathogens is essential. Therefore, the aim of the present paper is to discuss the interaction of infectious microorganisms with the immune system and show how knowledge of this interaction is being used to create better strategies for immunizations against pathogens. At the same time, the possibilities and limitations of vaccination procedures will be evident.

2. Interactions of the Immune System with Infectious Organisms

2.1 TWO TYPES OF IMMUNE RESPONSES

In general, two main types of active or acquired immune responses may occur: humoral and cell-mediated [5, 6]. A humoral immune response involves the production of antibodies to immunogenic components (usually proteins or polysaccharides), collectively called antigens. In this process, a particular area of an antigen (called an epitope) will bind very specifically with receptors (preformed antibody molecules) on an antibody-producing cell (B lymphocyte), which normally resides in lymphoid tissues. These lymphocytes are called B cells because they develop primarily in the bone marrow. Interaction with the antigen activates the cell to proliferate, differentiate, and secrete antibody molecules with the same binding specificity as those on its surface, which reacted with the antigen in the first place. In this way, antibodies of exquisite specificity for particular antigenic epitopes are produced and find their way into the serum of the individual. The antibodies are thus carried in the serum and as such belong to the humoral compartment.

The components of the acquired cell-mediated responses represent different subsets of T lymphocytes. These are called T cells because they develop primarily in the thymus. T cells do not secrete antibodies, but they possess mole-cules on their surface (called T cell receptors) that are similar to antibodies in their binding specificity for antigen epitopes. However, whereas antibodies can recognize and bind antigen alone, the T cell receptors cannot recognize antigen alone; the antigen fragment that they can interact with specifically can only be recognized if it is presented to the T cell in association with other molecules called major histocompatibility complex molecules or MHC-molecules. In this respect, it is said that the antigen fragments are presented to the T cells by the MHC molecules. T cells therefore have a double recognition requirement: for-eign antigen and self MHC molecules. There are mainly two types of T cells, helper (Th) and cytotoxic (Tc). Just as the name implies, Th cells assist other cells (such as B cells, Tc cells and macrophages, a type of phagocytic cell) to carry out their effector functions. Th cells accomplish this by providing acti-vating signals through direct interaction with the other cells and by secreting substances (called cytokines) that have various effects on the interacting cells. Tc cells on the other hand can under certain conditions attack, damage and kill cells that have become invaded by pathogens.

Th cells and Tc cells have other molecules on their surface that are essential for determining their interaction with antigen fragments. In the normal scheme of activities, Th cells have molecules called CD4, while Tc cells have molecules called CD8. CD4 molecules assure that Th cells interact with MHC molecules of class II, which are found on only a few types of cells in the body, mainly B cells, macrophages and dendritic cells. These cells can present antigen to Th cells in complex with MHC II molecules. CD8 molecules assure that Tc cells interact with MHC molecules of class I, that are found on all cells of the body that possess a nucleus. This means, practically speaking, that all cells can pres-ent antigen to Tc cells in complex with MHC I molecules (Figure 1, see end of chapter). Normally, Th cells do not react with antigen-MHC I complexes and Tc do not react with antigen-MHC II complexes.

2.2 TWO TYPES OF ANTIGENS ARE PROCESSED IN DIFFERENT WAYS

Whether an antigen gets presented to Tc cells by MHC I molecules or to Th cells by MHC II molecules is determined by the interaction of pathogenic mi-croorganisms with cells of the body. This is further determined by the way in which the cells take up the microorganisms and process their antigens as well as the route of trafficking of MHC I and MHC II molecules. According to this scheme, two types of antigens can be differentiated: exogenous and endo-genous.

Exogenous antigens (Figure 2, right side, see end of chapter) are taken up by cells by the process of endocytosis (or phagocytosis in the case of uptake by phagocytes). This means the antigens are internalized from the surface of the cells in membrane-bound vesicles, called endocytic (or phagocytic) vesicles.

These vesicles acquire hydrolytic enzymes that can degrade the antigens into small fragments. MHC II molecules are made within the endoplasmatic reticulum compartment of the cell and are targeted to endocytic vesicles, where antigen fragments are loaded into the antigen binding groove of the MHC II molecules, and these molecular complexes are transported to the surface of the antigen-presenting cell.

Endogenous antigens (Figure 2, left side, see end of chapter) are those that are from a foreign source, but are synthesized by the cells of the body. This situation arises primarily through the infection of cells by viruses. When this occurs, the genome of the virus is released into the cytoplasm of the cell and directs the cell to synthesize viral proteins. Some of the protein gets degraded by special enzyme complexes (the proteasome) in the cytoplasm, and these fragments are transported through the endoplasmatic reticulum membrane by special transporters (TAP), where they are loaded onto MHC I molecules. These complexes are then transported through the golgi apparatus to the cell surface, where they can be recognized by Tc cells. Tc cells can therefore recognize a cell that has been infected by a virus and can attack and kill that cell. In this way, the reproductive factory of the virus is effectively removed, preventing further production of the virus by that cell.

Usually, MHC I molecules do not cross the exogenous antigen route (they are not targeted to endocytic vesicles) and MHC II molecules are protected by the so-called invariant chain from antigen loading until they reach the endocytic compartments. This keeps the two types of antigens, endogenous and exogenous, separately distributed on MHC I and MHC II molecules, respectively.

2.3 DIFFERENT EFFECTOR FUNCTIONS FOR T HELPER AND T CYTOTOXIC CELLS

It was noted above that Tc cells can effect the elimination of virus-infected cells. Viruses are not the only important pathogens that can be fought by Tc cells. Although most bacterial infectious agents that are invasive enter cells through the endocytic-phagocytic pathway, some are able to break out of phagocytic vesicles into the cytoplasm of the host cell. Classical examples of such bacteria are *Listeria moncytogenes*, *Shigella flexneri* and *Rickettsia sp.* It is further known that many facultative intracellular infectious organisms such as *Salmonella typhi*, *Mycobacterium tuberculosis*, *Leishmania sp.*, *Legionella pneumophila*, as well as *Chlamydia trachomatis* can survive in phagocytic compartments for long periods of time. After a while, the vesicles become leaky, and some protein components of the microorganisms are released into the cytoplasm. In these cases, the proteins become degraded by proteasomes, the fragments enter the endoplasmatic reticulum and are loaded onto MHC I molecules. In this respect, Tc cells can be activated to many types of infectious organisms [4].

Th cells on the other hand do not attack infected cells, although there are exceptions to this rule. The assistance of Th cells is required, however, for full activation of B cells, T cells, and macrophages. Although they can not recognize antigen specifically, macrophages are extremely important as phagocytes of microorganisms. Some pathogens are able to resist killing and digestion by macrophages, and can indeed survive and multiply within these cells, as discussed above. Th cells can be divided into two subsets, Th1 and Th2. Macrophages can eliminate ingested microorganisms more efficiently if they have been activated by cytokines produced by the Th1 subset of Th cells. On the other hand, the cytokines produced by Th2 cells are important for the activation and differentiation of B cells to antibody-producing cells. Apparently, different antigens can preferentially induce one or the other Th cell subsets. Differential effects on immune responses due to the participation of different Th cells have been demonstrated for responses to the parasitic helminth *Schistosoma mansoni*, which were characterized by downregulation of Th1 cytokine production accompanied by induction of Th2 responses [7]. Also, the adjuvant effect of cholera toxin on the responses to tetanus toxoid in mice led to the induction of predominantly Th2 cells [8]. In contrast, recombinant *Salmonella* expressing fragment C of tetanus toxoid administered orally to mice elicited predominantly Th1-type responses [9].

From the above discussion, it should be gathered that the route of antigen uptake and processing is important for the kinds of responses elicited, which determine what components of the immune system will be armed for countering an infectious agent.

2.3.1 Humoral vs. Cell-mediated Immune Responses for Countering Infectious Agents

Antibodies are very useful in neutralizing toxins or in preventing the adherence to and colonization of tissues by some microorganisms. In the latter case, invasion of the tissues may thus be prevented, as in the example of microorganisms entering the body over mucosal surfaces. Also, entry of viruses into their target cells, which is mediated by structures on the surface of the viruses interacting with receptors on the target cells, can also be prevented by antibodies to the viral surface structures.

Nevertheless, it has been observed experimentally that cell-mediated immune responses are most protective in the case of infectious agents that are intracellular pathogens, that is, those that can enter, survive and multiply within cells. This would apply to viruses and rickettsias as well as to other bacteria that can survive within phagocytes. Antibodies can bind specifically to cell surface structures of microorganisms and aid in their uptake by professional phagocytes such as neutrophiles and macrophages, which then try to kill and degrade the microorganisms. As discussed above, however, some microorganisms are not readily killed by phagocytes, but can survive and even multiply intracellularly. Examples of such microorganisms are *Listeria moncytogenes*, *Shigella flexneri*,

Salmonella typhi, Mycobacterium tuberculosis, Legionella pneumophila, and Bacillus anthracis. In this case, antibodies are not very useful and cell-mediated immune responses are needed to combat the infectious process and spread of the microorganisms within the cellular compartment.

Many efforts in the past were directed at producing vaccines that could evoke high antibody titers. For controlling diseases such as tetanus, diphtheria and whooping cough, where protection depends primarily on the neutralization of toxins by antibodies, these immunizations have been highly successful. Protection from the majority of infectious diseases depends to a great extent, however, on the elicitation of strong cell-mediated immune responses. Indeed, the success of the first human vaccination experiment by Jenner against smallpox over 200 years ago was due primarily to the fact that a T cell immune response was elicited [10]. One of the challenges in vaccine development today is to prepare vaccines that can evoke the desired type of response.

2.3.2 Prophylactic vs. Therapeutic Vaccines

Another consideration that has to be made in vaccine development strategy is whether a prophylactic or a therapeutic vaccine is desired. A prophylactic vaccine is designed to prevent an initial infection by a microorganism or to prevent a toxin from exerting its effects, before the microorganism has had a chance to establish itself or any encounter with the toxin has occurred. In many cases, the elicitation of neutralizing antibodies may be effective, especially at mucosal surfaces where most infectious agents enter the body. With intracellular pathogens, however, vaccines that also elicit cell-mediated immune responses may be required, in case some of the microorganisms escape the neutralizing effects of antibodies and invade the tissues.

On the other hand, a therapeutic vaccine is designed to combat an established infection. There is a great deal of interest today, for example, in developing therapeutic vaccines to counter established infections caused by such agents as the AIDS virus or hepatitis B virus. A therapeutic vaccine should elicit antibodies to neutralize extracellular microorganisms or toxins that may be produced, but in the case of invasive pathogens it should also evoke cell-mediated responses to counter agents that have already entered cells. With regard to defense against biological warfare agents, the development of good therapeutic vaccines should be of particular interest, given the possibility of unanticipated attacks and the difficulties involved with immunizing entire civilian populations before an attack.

3. Types of Vaccines

A survey of the literature over the past few years reveals an enormous amount of activity in the area of the development of vaccines using biotechnologies, particularly genetic engineering. The goal of this research is the production of vaccines in forms that are safer for use (cause fewer side effects) and more effective than vaccines produced by conventional methods. Three basic types of vaccines can be produced using gene modification techniques: recombinant DNA-derived subunit vaccines, recombinant DNA inserted into heterologous carrier microorganisms [3,4,10,11], and „nude" DNA vaccines [12,13,14,15].

3.1 SUBUNIT VACCINES

Subunit vaccines consist of proteins or peptides (subunits) from disease-causing microorganisms. Immune responses to these proteins afford protection, and the proteins are often referred to as protective antigens. The genes directing the biosynthesis of these proteins are transferred to and cloned in non-pathogenic microorganisms such as the bacterium *Escherichia coli* or the yeast *Saccharomyces cerevesiae*, which subsequently produce these proteins in amounts which can be purified and used as vaccines. One successful vaccine that has been produced by this method is the one protecting against hepatitis B virus infection. A vaccine against hepatitis B was difficult to obtain because the virus causing this disease (fulminant hepatitis, chronic liver disease, liver cirrhosis and hepatocellular carcinoma) could not be grown in cell culture. However, the serum of infected individuals contains relatively large (22 nm in diameter) particles of a surface antigen (HBsAG) of the virus, which can be purified and used as a vaccine. While this vaccine is very effective, the source is limited, so the HBsAG was a prime target for subunit vaccine development using genetic engineering [16]. The first attempts to clone the gene encoding the HBsAG in *Escherichia coli* resulted in suboptimal expression, but the baker's yeast *Saccharomyces cerevesiae* could produce relatively large quantities of the protective antigen after transfection with a plasmid vector containing the gene and an expression system that could direct the production of the protein in the yeast. When isolated, the protein has the property that it aggregates into particles similar to those found in serum. These particles are apparently especially immunogenic, that is, they induce strong immune responses. Other attempts at producing subunit vaccines have been less successful, at least as far as efficacy of the vaccines is concerned [17], most probably because the proteins do not aggregate into large particles that are strongly immunogenic.

One problem with subunit vaccine production, such as in the HBsAG system, is that the proteins produced are not exported into the medium, and they have to be extracted from the cells and purified. To solve this problem, a relatively new innovation is to incorporate an efficient bacterial protein export system into the vector construct [18]. Another problem with subunit vaccines, is that they would tend to elicit only Th and not Tc responses.

3.1.1 Improvement of the Immunogenicity of Subunit Vaccines with Adjuvants
One way to enhance the immunogenicity of proteins is to mix it with substances

called adjuvants (from Latin *adjuvare*, to help). Adjuvants can be divided into two major families: vehicles and immunomodulators. Vehicles include liposomes, emulsions, and substances called immunostimulating complexes (IS-COMS), that help to transport antigens to and retain them in, lymphoid tissues in order to improve the immune response [19]. Immunomodulators include substances that stimulate cells of the immune system in an antigen-nonspecific manner to proliferate and produce cytokines. Some adjuvants of course have both properties. An adjuvant that has been used frequently in the past is aluminum potassium sulfate [5]. Several saponins, substances extracted from the plant *Quillaja saponaria*, are currently being tested in clinical trials [19]. One saponin, QS-21, elicits mixed Th1 and Th2 responses, and will soon enter human trials. SmithKline Beecham Biologicals is currently testing a combination of this saponin with a microfluidized squalene emulsion and another adjuvant, monophosphoryl lipid A (MPL) [19]. A further method for improving immunogenicity of a protein is the microencapsulation of antigens into microscopic spheres using biodegradable substances such as polyesters. The aim of this work is to achieve a controlled release of the protein in the body for a better stimulus of the immune system, and induce protection with a single-shot vaccine [20]. Another strategy is to fuse the protein antigen by genetic engineering methods to a second protein that is known for its ability to aggregate and to induce strong immunogenic responses [21].

3.1.1.1 Mucosal Immunity. As mentioned in the discussion on adjuvants, one goal is to target antigens to particular tissues, where the immune responses will be most effective. In this regard, the mucosal tissues play a special role, as most infections begin at mucosal sites [22, 23]. Also, when considering delivery systems for biological weapons, the aerosol route is particularly effective for many pathogens [24], so that mucosal immunity would be highly desired in countering such attacks. There has been a great deal of work in the past few years on how to target vaccines to mucosal sites. Several substances, when combined with antigens, have been shown to be effective in this respect, and have been termed mucosal adjuvants. Among these are cholera toxin [25], a genetically inactivated subunit derived from the heat labile toxin of the bacterium Escherichia coli [19], and the fibronectin-binding protein I of the bacterium Streptococcus pyogenes [26]. Because the mucosal system is widely distributed in the body, it would seem difficult to achieve immunity throughout this vast system. However, it has been shown that immunization in one mucosal site apparently affords protection in other mucosal sites, possibly due to the fact that immune cells of the mucosal system in one site can wander and home into distant sites [19]. One problem that arises frequently when trying to achieve mucosal immunity by immunization is the development of systemic tolerance to the antigen. This means that immune responses to the applied antigen in parts of the body other than the mucosal areas are repressed. However, ways of overcoming this effect are being developed. Apparently, the use of particular Gram-positive bacteria to target antigens to mucosal surfaces results in both mucosal and systemic responses to the antigen (see 3.2.2, Bacterial Vectors) [22, 23].

Even with the use of adjuvants, even those directing antigens to mucosal areas, the type of immune response to the protein antigens is still a problem. Proteins administered in this way would produce primarily antibody and Th responses, but not Tc responses. Indeed, it has been shown that cholera toxin used as a mucosal adjuvant effects the activation of mainly Th2 cells [25], which would help boost antibody production but not cell-mediated immunity to the antigens of the pathogen.

3.2 RECOMBINANT DNA INSERTED INTO HETEROLOGOUS CARRIER MICROORGANISMS

3.2.1 Viral Vectors

A further method of vaccine development is to insert protective antigen genes into heterologous carriers. In this regard, several viral vectors have been developed to carry foreign genes. Since these viral vectors would deliver the genes to the cytoplasm of the cells they invade, antibody and both Th and Tc responses would be stimulated.

The genetic manipulation of vaccinia virus, the virus used for smallpox vaccination, will be given as an example of such manipulations (Figure 3, see end of chapter), although it must be emphasized that it is difficult to generalize because of the different types of viruses and various schemes of viral multiplication. In the case of vaccinia, the gene that is to be transferred is built into a plasmid (recombination vector). The gene is constructed to contain short pieces of DNA homologous to the vaccinia genome flanking both of its ends, thus assuring that the new gene will be integrated into the viral genome by homologous recombination. The plasmid is artificially transfected into host cells *in vitro*, and the virus is allowed to infect these cells. Inside the host cell, a homologous recombination of the new gene with the vaccinia genome will occur [27]. Thereafter, the virus carries the new gene, which will be expressed in the host cell along with the expression of the vaccinia genome. At this point it might be mentioned that vaccinia virus has a large genome that will allow insertion of up to 20 kilobases of DNA. Furthermore, it induces a strong immune response in the host. Therefore, it was perceived as a likely carrier of multiple foreign antigen genes whose products, when expressed in the host, would lead to effective immune responses. It was speculated that protection against multiple pathogens could be achieved in this way [28].

The rationale for this approach is the presumption that these carrier microorganisms will invade cells, serve as a long term source of antigen, and provoke both Th and Tc responses in the vaccinated individual. However, most adults over the age of 21 today have had smallpox vaccinations, and it has been observed that such individuals carry enough antibodies to neutralize vaccinia virus before it can get established and direct the expression of its genes [29], so that its use as a vector in these individuals is apparently limited.

There has been a great deal of work in recent years on the possibility of using adenoviruses as gene vectors. Adenoviruses, the causative agents of acute respiratory and ocular infections, are widespread in nature. They are among the most efficient gene delivery vehicles in use today. They have a broad host range and do not integrate into chromosomal DNA, which reduces the risk of mutations via insertion. Retroviruses, by contrast, integrate randomly into the genome of the host. Furthermore, adenoviruses can be produced at high titers (up to 10^{10} per milliliter) [30, 31]. Until recently, there were problems associated with their protein coat antigens causing immune reactions that would clear the body of the vectors before they might have a chance to be expressed. This has apparently been solved by construction of helper-dependent adenoviruses, in which all viral coding sequences have been replaced with foreign genes, so that viral protein expression (and an immune response against them) is eliminated. The modified viruses also have a carrying capacity of up to 40 kb of insert DNA. It was shown that these viruses could still infect cells just as efficiently as the wild type viruses, and the foreign genes they were carrying were expressed in the host cells [31]. Alternatively, the development of adeno-associated viruses as vectors for gene delivery seems promising, as these viruses are defective by nature and have thus never been shown to have any pathogenic effects in humans [32].

Another delivery system currently under investigation uses alphaviruses as vectors. Although several alphaviruses are pathogenic, the ones used as vectors, Semliki Forest virus and Sindbis virus, are usually considered avirulent for humans [33]. However, natural variants of these viruses have been found to cause painful human disease. Sindbis virus strains have been found to cause polyarthritis or a disease characterized by fever and rash, while Semliki Forest virus strains were reported to cause severe headache, fever, myalgia and arthralgia [34], so that they must be considered as potential pathogens in some cases. As with the adenoviruses, helper-dependent alphavirus vectors have been constructed by removing the part of the virus genome encoding the protein coat antigens. Theoretically, these viruses are not able to replicate, but there are situations where reversion of the virus to a replicative form can occur, so that biosafety concerns require refinement of the system [33, 34].

Several other viruses are being investigated as potential recombinant vaccines (Table 1, see end of chapter). Recombinant polioviruses are of particular interest, as they are known to be safe and elicit long-lasting immunity [35]. Some viruses are being used as carriers of multiple minigenes, which are fragments of genes encoding just antigenic epitope portions of proteins, cloned and expressed as one polyepitope (Figure 4, see end of chapter). It has been shown, for example, that B, Th, and Tc cell responses directed against the appropriate epitopes were mounted in mice infected with the recombinant viruses [36, 37].

3.2.2 Bacterial Vectors

The bacterium *Salmonella typhi* (the causative agent of typhoid fever) is an organism that can survive and even multiply inside macrophages, one of the major phagocytes of the body. The interest in *Salmonella sp.* as a carrier lies in the fact that this organism can target antigens to the gut-associated lymphoid tissue (mucosal site) where different kinds of responses including secretory, humoral and cellular immune responses (including activation of both Th and Tc cells) are induced by the antigen (see also the discussion in section 2.3 above). Some *Salmonella* mutants are attenuated in their virulence, and are being developed as vectors for antigen genes [38, 39]. One disadvantage that might be connected with their use are the potential negative side effects that could occur due to the possession of endotoxin as a cell wall component, a highly active substance biologically that may cause severe chills and fever.

In this respect, some non-pathogenic, commensal Gram-positive bacteria such as *Lactococcus lactis* and *Streptococcus gordonii*, which do not contain endotoxin, are being developed as vectors for antigens [22, 40]. While it has been shown that these Gram-positive bacteria are indeed targeted to mucosal sites and induce mucosal-associated antibodies, it is not clear whether cell mediated immunity is also elicited. In some cases, however, production of local mucosal antibody as well as systemic antibody responses using *Streptococcus gordonii* as a vector has been reported [22, 23]. The systemic antibodies were of the IgG type [23], which would suggest that Th2 T cells were actively involved in their production.

Another approach with Gram-positive bacteria as vectors exploited the special function of listeriolysin O (LLO), a substance produced by pathogenic strains of the bacterium *Listeria monocytogenes*. As discussed above, this bacterium is taken up by macrophages into the cytoplasm in a membrane-bound vesicle called a phagosome. Inside the phagosome, the bacterium produces LLO, which can disrupt the vesicle membrane, releasing the organisms into the cytoplasm of the phagocyte. When they are free in the cytoplasm, the proteins they produce are processed by the endogenous route, leading to the induction of a cell-mediated, Tc response. When mice were injected with a non-pathogenic, vaccine strain of *Bacillus anthracis* carrying the gene for LLO, they mounted a Tc cell response directed against LLO, which protected them from a challenge infection with *Listeria monocytogenes* [41]. This showed that the *Bacillus anthracis* organisms produced LLO in vivo, which was released into the cytoplasm of infected cells. Thus, *Bacillus anthracis* acted as a vector for delivering LLO to the appropriate compartment for endogenous antigen processing. This also raises the possibility of using this engineered form of vaccine to deliver *Bacillus anthracis* antigens to the cell cytoplasm for endogenous antigen processing. In this way, Tc immune responses might be elicited against phagocytes containing intracellular *Bacillus anthracis* organisms. This would be advantageous in combatting anthrax aquired as a result of inhalation of spores, as the organisms readily become sequestered within macrophages, where they are protected from antibodies as well as antibiotics and can spread throughout the lymphatic system.

3.2.3 Plant Vectors

It should be mentioned that investigators at the Boyce Thompson Institute for Plant Research are introducing genes encoding antigens from pathogens into plants, such as bananas and potatoes. According to a member of the research team, mice fed the raw potatoes developed immune responses to those antigens [42].

3.3 DNA VACCINES

One of the most exciting developments in the past few years has been the use of DNA as vaccines [43]. These vaccines consist of small, non-replicating plasmid DNA expression vectors encoding an antigen of interest (Figure 5, see end of chapter). They are administered by injection into skeletal muscles or into the skin by bombardment with DNA-coated gold particles using a gene gun. The antigens are expressed in host cells, provoking an immune response consisting of both antibody and cell-mediated components. Some models that have been investigated are presented in Table 2 (see end of chapter). One of the great surprises of this technique was how good it functioned, initiating a „new era in vaccines and immune therapeutics" [12]. Minute (nanogram-microgram) amounts are sufficient to elicit an immune response that in experimental models lasted for over a year. Interestingly, it has been determined that bacterial plasmid DNA contains immunostimulatory DNA sequences (ISS) that function as adjuvants (Figure 5, see end of chapter) [13]. These sequences (CpG motifs) are found far more frequently in bacterial and viral genomes than in vertebrate genomes, and may account for the great efficiency of DNA vaccines to elicit immune responses. In several types of animal models, DNA vaccines have induced antibodies and cell-mediated (Tc and Th1) responses [12, 43], whereas protein immunization elicits primarily antibodies and Th2 cells, as has already been mentioned. In effect, DNA vaccines have the strong features of live, attenuated viral vaccines while affording additional safety features. Furthermore, generic production and verification techniques can be used to manufacture a spectrum of DNA vaccines, which can greatly reduce the time and cost of development and production [15].

Despite all these advantages, DNA vaccines have several limitations [15]. One is that they can only target protein components of microorganisms for protective responses. For several microorganisms, polysaccharide surface components represent the protective antigens of choice. Furthermore, DNA immunizations do not lead to the induction of mucosal type responses. Future modifications might include microencapsulation for more efficient uptake into cells, or the inclusion of proteins that can facilitate targeting to specific tissues, such as mucosal tissue [43]. In one modification of the technique, multiple antigen epitopes in the form of one polyepitope were incorporated into a DNA vaccine. Mice that were vaccinated with this preparation responded by producing MHC I-restricted Tc responses to each of the epitopes [44].

4. Summary

In the past, vaccine development was concerned mainly with enhancing antibody titers to antigens of pathogenic microorganisms. The importance of cell-mediated immune responses in combating most pathogens has, however become quite evident. Depending on the type of vaccine and how it is administered, very different types of immune responses including humoral, Th1, Th2, or Tc cell responses may be elicited. Above all, the ways in which antigens are taken up and processed by antigen-presenting cells must be considered in designing a vaccination strategy. Although vaccination strategies capable of inducing the desired humoral and cell-mediated responses in humans have not been easy to achieve, some developments in the last few years have had a particularly positive impact. The elucidation of the ways in which antigens are processed and presented to cells of the immune system, as well as the development of special antigen vectors through the techniques of genetic engineering will greatly aid future efforts. One particularly exciting development that shows much promise is the use of DNA vaccines. Vaccines of course do not afford a perfect defense against biological warfare agents. Immunizations have to be given in advance of an attack, the protection achieved is often less than one hundred percent, and there are problems in immunizing large groups, such as civilian populations (or even troops), in the event of aggression. Still, vaccines can provide effective protection against many infectious diseases, and the possession of a good vaccine against a particular agent can act as a deterrent to discourage an aggressor intending to use that agent.

5. References

1. DiRita, V.J. (1994) Multiple regulatory systems in *Vibrio cholerae* pathogenesis, *Trends Microbiol.* **2**, 37-38.
2. Straley, S.C. and Perry, R.D. (1995) Environmental modulation of gene expression and pathogenesis in *Yersinia, Trends Microbiol.* **3**, 310-317.
3. Leclerc, C., and Ronco, J. (1998) New approaches in vaccine development, *Immunol. Today* **19**, 300-302.
4. Hess, J., and Kaufmann, S.H.E. (1996) Neue Strategien in der Impfstoff-Entwicklung, *Biospektrum* **4**, 18-23.

174

5.   Kuby, J. (1997) Immunology, Third Edition, W.H. Freeman and Company, New York.
6.   Abbas, A.K., Lichtman, A.H., Pober, J.S. (1997) *Cellular and Molecular Immunology*, Third Edition, W.B. Saunders Company, Philadelphia.
7.   Pearce, E.J., Caspar, P., Grzych, J.-M., Lewis, F.A., and Sher, A. (1991) Downregulation of Th1 cytokine production accompanies induction of Th2 responses by a parasitic helminth, *Shistosoma mansoni, J. Exp. Med.* **173**, 159-166.
8.   Marinaro, M., Staats, H.F., Hiroi, T., Jackson, R.J., Coste, M., Boyaka, P.N., Okahashi, N., Yamamoto, M., Kiyono, H., Bluethmann, H., Fujihashi, K., and McGhee, J.R. (1995) Mucosal adjuvant effect of cholera toxin in mice results from induction of T helper 2 (Th2) cells and IL-4, *J. Immunol.* **155**, 4621-4629.
9.   VanCott, J.L., Staats, H.F., Pascual, D.W., Roberts, M., Chatfield, S.N., Yamamoto, M., Coste, M., Carter, P.B., Kiyono, H., and McGhee, J.R. (1996) Regulation of mucosal and systemic antibody responses by T helper cell subsets, macrophages, and derived cytokines following oral immunization with live recombinant *Salmonella, J. Immunol.* **156**, 1504-1514.
10. Bona, C.A., Casares, S., and Brumeanu, T.-D. (1998) Towards development of T-cell vaccines, *Immunol. Today* **19**, 126-133.
11. Liew, F.Y. (1988) Biotechnological trends towards synthetic vaccines, *Immunol. Lett.* **19**, 241-244.
12. Chattergoon, M., Boyer, J., and Weiner, D.B. (1997) Genetic immunization: a new era in vaccines and immune therapeutics, *FASEB J.* 11, 753-763.
13. Tighe, H., Corr, M., Roman, M., and Raz, E. (1998) Gene vaccination: plasmid DNA is more than just a blueprint, *Immunol. Today* **19**, 89-97.
14. Ulmer, J.B., Donnelly, J.J., and Liu, M.A. (1996) DNA vaccines promising: a new approach to inducing protective immunity, *ASM News* **62**, 476-479.
15. Young, P. (1997) Bright outlook on direct DNA immunizations, *ASM News* **63**, 659-663.
16. Stephenne, A. (1988) Recombinant versus plasma-derived hepatitis B vaccines: issues of safety, immunogenicity and cost-effectiveness, *Vaccine* **6**, 299-303.
17. Welkos, S.L., and Friedlander, A.M. (1988) Comparative safety and efficacy against *Bacillus anthracis* of protective antigen and live vaccines in mice, *Microb. Pathog.* **5**, 127-139.
18. Tzschaschel, B.D., Guzman, C., Timmis, K.N., and de Lorenzo, V. (1996) An *Escherichia coli* hemolysin transport system-based vector for the export of polypeptides: Export of Shiga-like toxin IIeB subunit by *Salmonella typhimurium* aroA, *Nature Biotechnol.* **14**, 765-769.
19. Van Regenmortel, M. (1997) Searching for safer, more potent, better-targeted adjuvants, *ASM News* **63**, 136-139.

20. Kissel, T., Koneberg, R., Hilbert, A.K., and Hungerer, K.-D. (1997) Micro-encapsulation of antigens using biodegradable polyesters: facts and phantasies, *Behring Institute Mitteilungen* **98**, 172-183.

21. Hancock, R.E.W., and Wong, R. (1997) Potential of protein OprF of *Pseudomonas* in bivalent vaccines, *Behring Institute Mitteilungen* **98**, 283-290.

22. Fischetti, V.A. (1996) Gram-positive commensal bacteria deliver antigens to elicit mucosal and systemic immunity, *ASM News* **62**, 405-410.

23. Medaglini, D., Pozzi, G., King, T.P., and Fischetti, V.A. (1995) Mucosal and systemic immune responses to a recombinant protein expressed on the surface of the oral commensal bacterium *Streptococcus gordonii* after oral immunization. *Proc. Natl. Acad. Sci. USA* **92**, 6868-6872.

24. Bundesamt für Zivilschutz .(1975) *Zivilschutz. Gefahren aus der Retorte. Gesundheitliche Aspekte bei chemischen und biologischen Kampfmittel,* Band 6, BZS-Schriftenreihe, Weltgesundheitsorganisation (WHO), Mönch, Bonn].

25. Cong, Y., Weaver, C.T., and Elson, C.O. (1997) The mucosal adjuvanticity of cholera toxin involves enhancement of costimulatory activity by selective up-regulation of B7.2 expression, *J. Immunol.* **159**, 5301-5308.

26. Medina, E., Talay, S.R., Chhatwal, G.S., and Guzman, C.A. (1998) Fibronectin-binding protein I of *Streptococcus pyogenes* is a promising adjuvant for antigens delivered by mucosal route, *Eur. J. Immunol.* **28**, 1069-1077.

27. Moss, B. (1985) Vaccinia virus expression vector: a new tool for immunologists, *Immunol. Today* **6**, 243-245.

28. Perkus, M.E., Piccini, A., Lipinskas, B.R., and Paoletti, E. (1985) Recombinant vaccinia virus: immunization against multiple pathogens, *Science* **229**, 981-984.

29. Restifo, N.P. (1996) The new vaccines: building viruses that elicit anti-tumor immunity, *Current Opin. Immunol.* **8**, 658-663.

30. Morsy, M.A., and Caskey, C.T. (1997) Safe gene vectors made simpler, *Nature Biotechnol.* **15**, 17.

31. Kochanek, S., Clemens, P.R., Mitani, K., Chen, H.-H., Chan, S., and Caskey, C.T. (1996) A new adenoviral vector: Replacement of all viral coding sequences with 28 kb of DNA independently expressing both full-length dystrophin and β-galactosidase, *Proc. Natl. Acad. Sci. USA* **93**, 5731-5736.

32. Carter, B.J. (1996) The promise of adeno-associated virus vectors, *Nature Biotechnol.* **14**, 1725-1725.

33. Berglund, P., Tubulekas, I., and Liljeström, P. (1996) Alphaviruses as vectors for gene delivery, *Trends Biotechnol.* **14**, 130-134.

34. Strauss, J.H., and Strauss, E. (1994) The alphaviruses: gene expression, replication, and evolution, *Microbiol. Rev.* **58**, 491-562.

35. Rolph, M.S., and Ramshaw, I.A. (1997) Recombinant viruses as vaccines and immunological tools, *Current Opin. Immunol.* **9**, 517-524.

176

36. Thomson, S.A., Elliott, S.L., Sherritt, M.A., Sproat, K.W., Coupar, B.E.H., Scalzo, A.A., Forbes, C.A.; Ladhams, A.M., Mo. X.Y., Tripp, R.A., Doherty, P.C., Moss, D.J., and Suhrbier, A. (1996) Recombinant polyepitope vaccines for the delivery of multiple CD8 cytotoxic T cell epitopes, *J. Immunol.* **157**, 822-826.
37. An, L., and Whitton, J.L. (1997) A multivalent minigene vaccine, containing B-cell, cytotoxic T-lymphocyte, and Th epitopes from several microbes, induces appropriate responses in vivo and confers protection against more than one pathogen, *J. Virol.* **71**, 2292-2302.
38. Dougan, G., Chatfield, S., Pickard, D., Bester, J., O'Callaghan, D., and Maskell, D. (1988) Construction and characterization of vaccine strains of *Salmonella* harboring mutations in two different *aro* genes, *J. Infect. Dis.* **158**, 1329-1335.
39. Levine, M.M., Galen, J., Barry, E., Noriega, F., Tacket, C., Sztein, M., Chatfield, S., Dougan, G., Losonsky, G., and Kotloff, K. (1997) Attenuated *Salmonella typhi* and *Shigella* as live oral vaccines and as live vectors, *Behring Institute Mitteilungen* **98**, 120-123.
40. Steidler, L., Robinson, K., Chamberlain, L., Schofeld, K.M., Remaut, E., Le Page, R.W.F., and Wells, J.M. (1998) Mucosal delivery of murine interleukin-2 (IL-2) and IL-6 by recombinant strains of *Lactococcus lactis* co-expressing antigen and cytokine, *Infect. Immun.* **66**, 3183-3189.
41. Sirard, J.-C., Fayolle, C., de Chastellier, C., Mock, M., Leclerc, C., and Berche, P. (1997) Intracytoplasmic delivery of Listeriolysin O by a vaccinal strain of *Bacillus anthracis* induces CD8-mediated protection against *Listeria monocytogenes*. *J. Immunol.* **159**, 4435-4443.
42. Fox, J. (1997) Taking multiple paths toward mucosal immunity, *ASM News*, **63**, 413-414.
43. Ulmer, J.B., Sadoff, J.C., and Liu, M.A. (1996) DNA vaccines, *Curr. Opin. Immunol.* **8**, 531-536.
44. Thomson, S.A., Sherritt, M.A., Medveczky, J., Elliott, S.L., Moss, D.J., Fernando, G.J.P., Brown, L.E., and Suhrbier, A. (1998) Delivery of multiple CD8 cytotoxic T cell epitopes by DNA vaccination, *J. Immunol.* **160**, 1717-1723.

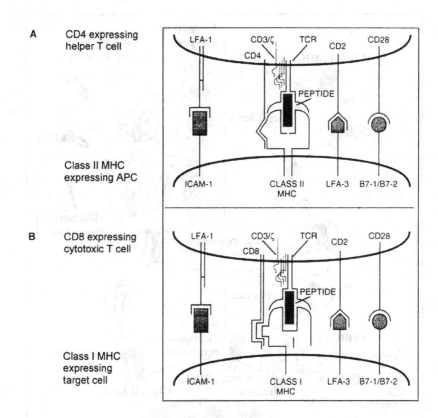

Figure 1. T lymphocyte surface molecules and their ligands involved in antigen recognition and T cell responses. Interactions between a CD4[+] helper T lymphocyte and an antigen presenting cell (APC) (A) or a CD8[+] cytotoxic T lymphocyte and a target cell (B) involve multiple T cell surface proteins that recognize different ligands on the APC or target cell. The T cells recognize antigen peptide fragments (peptide) in association with class II MHC molecules (CD4[+] helper T lymphocytes) or class I MHC molecules (CD8[+] cytotoxic T lymphocytes). LFA = leucocyte function-associated antigen; ICAM = intercellular adhesion molecule; MHC = major histocompatibility complex. Source: Ref. 6, used with kind permission from W.B. Saunders Company.

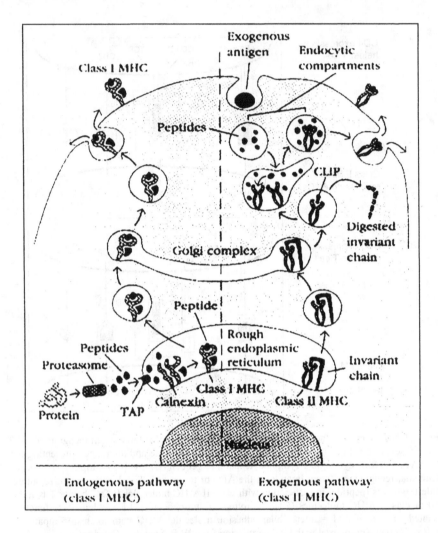

Figure 2. Model of separate antigen-presenting pathways for endogenous (left) and exogenous (right) antigens. The mode of antigen entry into cells and the site of antigen processing appear to determine whether antigenic peptides associate with class I MHC molecules in the rough endoplasmic reticulum or with class II molecules in endocytic compartments. Some elements of this model have not been experimentally demonstrated. <u>Source</u>: Ref. 5, Immunology 3/E by J. Kuby © 1997 by Janis Kuby. Used with kind permission of W.H. Freeman and Company.

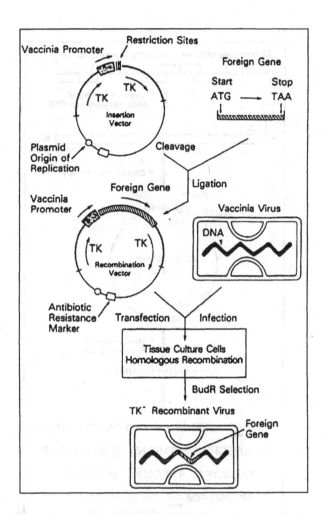

Figure 3. Construction of vaccinia virus recombinants that express a foreign gene. In the first step, the foreign gene that is to be transferred is built into a plasmid vector containing a vaccinia virus promoter (DNA region regulating synthesis), restriction enzyme sites for insertion of a foreign gene, flanking vaccinia virus DNA from the thymidine kinase (TK) gene that will specify the site of homologous recombination, an origin of replication and an antibiotic resistance gene for propagation in Escherichia coli. After insertion of the foreign gene, the plasmid is replicated in Escherichia coli to obtain sufficient numbers of the plasmid, and then used to artificially transfect tissue culture cells that are also infected with vaccinia virus. Homologous recombination introduces the the foreign gene into the vaccinia genome. Source: Ref. 27. Reprinted from Immunology Today, Vol. 6, Bernard Moss, pages 243-245 © 1985, with kind permission from Elsevier Science.

180

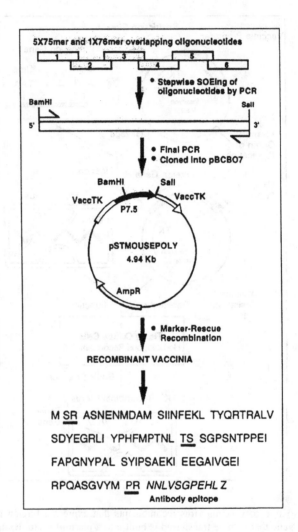

Figure 4. Construction of a recombinant vaccinia virus expressing the murine polytope protein. The DNA sequence coding for the polytope protein, which contains 10 epitopes recognized by Tc cells in association with MHC I molecules (Tc cell epitopes), was made by joining six overlapping oligonucleotides using the polymerase chain reaction technology. The full-length DNA fragment was inserted, via two plasmids, into vaccinia to generate the recombinant polytope vaccinia virus. The sequence in italics represents a linear B cell epitope from Plasmodium falciparum, and the underlined amino acids represent engineered restriction sites. Source: Ref. 36, used with kind permission from The Journal of Immunology.

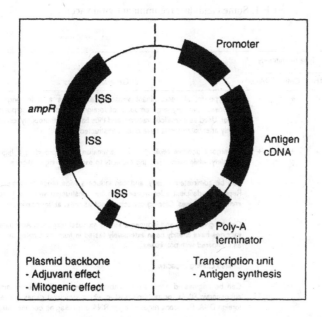

Figure 5. Plasmid DNA for gene vaccination has two major units: (1) a plasmid backbone that delivers adjuvant and mitogenic activity via immunostimulatory sequences (ISS); and (2) a transcription unit comprising a promoter, antigen cDNA and poly-adenylation (A) addition sequence, which together direct protein synthesis. ISS are shown schematically as being located within the ampicillin antibiotic resistance gene (ampR) and the noncoding region of the plasmid. Although antigen synthesis requires sequence integrity of the transcription unit, this is not the case for the adjuvant/mitogenic effects, since ISS-containing oligonucleotides exert the same effects as ISS-enriched noncoding plasmid DNAs. Source: Ref. 13. Reprinted from Immunology Today, Vol. 19, H. Tighe, M. Corr, M. Roman, and E. Ratz, Gene vaccination: plasmid DNA is more than just a blueprint, pages 89-97 © 1998, with kind permission from Elsevier Science.

TABLE 1. Some candidate recombinant viral vaccines

Virus type	Type of immunity			General comments
	Ab	CMI	Muc	
Vaccinia	+	+	+	Prototype for all recombinant viruses. Can induce a wide range of immune responses, and very large amounts of foreign DNA can be incorporated into the vector. Used as a smallpox vaccine, and thus has been thoroughly tested in humans. Highly attenuated forms have been constructed.
Avipox	+	+	+	Undergoes abortive replication in mammalian cells, providing a high degree of biosafety, while maintaining the capacity to express foreign antigens.
Adenovirus	+	+	+	Can be administered orally, and can induce a wide range of immune responses. Replication-deficient adenoviruses can be constructed which still induce strong immune responses. Cloning capacity is not as great as for poxviruses.
Polio	+		+	Can be given orally and induces a strong mucosal response. Attenuated poliovirus vaccines have already been extensively tested in humans. Limited cloning capacity as compared with poxviruses.
Mengo	+	+		Limited cloning capacity.
SFV	+	+		Can be engineered either as recombinant infectious RNA, expressing additional subgenomic RNAs, or as 'replicons' in which structural genes are replaced with foreign DNA. Replicons require helper RNA or packaging cell lines for production.
Sindbis	+	+		Similar characteristics to SFV. Genetic stability of a recombinant expressing immunogenic proteins from Japanese encephalitis virus was low.
VEE	+	+		There is little or no pre-existing immunity to this vector in human populations.
Influenza	+	+	+	Induction of strong mucosal immunity. Pre-existing vector-specific immunity could potentially be avoided by using vectors belonging to different influenza virus subtypes.

This table is a non-exhaustive list of types of recombinant viruses that have been tested as vaccines, showing the classes of immune response that have so far been demonstrated for each type. Ab, antibody; CMI, cell-mediated immunity; Muc, mucosal immunity; SFV, Semiliki Forest virus; VEE, Venezuelan equine encephalitis.

Source: Ref. 35, used with kind permission from Current Biology Ltd.

TABLE 2. Potential targets of DNA vaccines

Investigator(s)	Pathogen	Target	Delivery site	Animal model
Cox	BHV-1	Glycoprotein	Intramuscular	Mouse/cattle
Conry	CEA/tumor	CEA	Intramuscular	Mouse
Donnelly/Lui	CRPV	L1	Intramuscular	Rabbit
Davis	HBV	sAg	Intramuscular	Mouse
Purcell	HBV	sAg	Intramuscular	Chimpanzee
Ray	HCV	Core	Intramuscular	Mouse/rat
Boyer	HIV-1	*env* and *rev*	Intramuscular	Chimpanzee
Wang	HIV-1	*env* and *rev*	Intramuscular	Mouse/macaques
Agadjanyan	HIV-2	*env*	Intramuscular	Mouse/macaques
Chalian	HPV	E6	Intramuscular	Mouse
Rouse	HSV	Glycoprotein B	Intramuscular	Mouse
Agadjanyan	HTLV-1	*env*	Intramuscular	Rat/rabbit
Ulmer/Montgomery	Influenza B	Influenza B	Intramuscular	Mouse
Fynan/Robinson	Influenza virus	Hemagglutinin	Gene gun Mucosal Parenteral	Mouse
YankaucKas	Influenza virus	Hemagglutinin	Intramuscular/Intradermal	Mouse
Sedagah	Malaria	Circumsporozoite	Intramuscular	Mouse
Geissler	MHC-1	Glycoprotein	Intramuscular	Rat
Plautz	MHC-1	Glycoprotein	Tumor	Mouse
Herrmann, J.E.	Murine rotavirus	VP4, VP6 & VP7	Intramuscular	Mouse
Xiang	Rabies	Glycoprotein	Intramuscular	Mouse
Lu/Robinson/Letvin	SIV	*env* and *gag*	Gene gun	Macaques
Tascon/Lowie	Tuberculosis	hsp65	Intramuscular	Mouse
Huygen/Ulmer	Tuberculosis	Antigen 85	Intramuscular	Mouse

Source: Ref. 12, used with kind permission from *The Federation of American Societies for Experimental Biology (FASEB) Journal*.

THE DEVELOPMENT OF AN ORAL VACCINE AGAINST ANTHRAX

NETTY D. ZEGERS, EERENST KLUTER, HANS VAN DER
STAP, ERNST VAN DURA, PHILIP VAN DALEN, AND MIKE
SHAW
TNO Prevention and Health
Division of Immunological and Infectious Diseases
P.O.Box 2215, 2301 CE Leiden, The Netherlands

1. Introduction

1.1 ANTHRAX

Bacillus anthracis, a sporulating gram-positive rod, is the causative agent of the disease anthrax, which commonly occurs among herbivores but can also affect humans. The disease is still prevalent in many parts of Africa, Central and South America, the Indian subcontinent, Indonesia, Australia, the Middle East and certain regions of the former USSR. The disease in humans occurs primarily in a cutaneous form, but also as pulmonary or gastro-intestinal manifestations, depending upon the route of infection by anthrax spores (Table 1). In endemic countries anthrax occurs mainly as an occupational disease. Its normal spread among herbivores causes accidental infection to people whose occupation brings them into contact with animal products. For this reason anthrax is also called "wool sorter's disease".

TABLE 1. Forms of Anthrax

• Cutaneous
  ▪ Contact with animal products:
   hides, wool, (wool sorter's disease)
  ▪ 80% of patients recover
• Intestinal
  ▪ Consumption of infected meat
	• Inhalation
	  ▪ Inhalation of aerosols
	  ▪ 100% lethal unless antibiotic therapy
	   starts within 24 hours after exposure
	  ▪ Starts mildly: flu-like symptoms

A. Kelle et al. (eds.), The Role of Biotechnology in Countering BTW Agents, 185–196.
© 2001 *Kluwer Academic Publishers. Printed in the Netherlands.*

The cutaneous form is transmitted through a break in the skin and is manifested as a painless carbuncle or pustule. The infection may spread through the bloodstream and can cause shock, cyanosis, sweating and collapse. Without treatment 10-20% of all cases of cutaneous infection progress to fatal septicemia. With hospital treatment (antibiotics) this figure drops to 0-3%.

The gastro-intestinal form is transmitted by consumption of meat from animals that have died from anthrax. Symptoms are nausea, vomiting, fever, chills, abdominal pain, bloody diarrhea, general discomfort, malaise and shock. Finally, a comatose stage is reached, eventually leading to death. Gastro-intestinal anthrax is far more often fatal than the cutaneous form.

The pulmonary form of anthrax is transmitted by inhalation of spores in dust clouds created by handling dry hides, wool and bone meal. Airborne spores that are >5 micron in size pose a reduced threat, as they are either physically trapped in the nasopharynx or are cleared by the mucociliary system. Spores between 1 and 5 micron in size are more likely to penetrate and be deposited on alveolar ducts or alveoli. They are phagocytosed by alveolar macrophages and are transported to the mediastinal lymph nodes where they germinate and multiply rapidly, causing hemorrhagic mediastinitis. Illness begins insidiously with mild fever, fatigue and malaise. This mild initial phase is terminated by the sudden onset of acute illness. The rapid development of dyspnoea, cyanosis, severe pyrexia is quickly followed by coma and death. The fatality rate of the pulmonary form is or is near 100% [1].

Human anthrax is not considered contagious and there is no evidence of person-to-person transmission.

1.2 ANTHRAX TOXINS

Anthrax is caused by the three-part protein toxin secreted by *B. anthracis* (Table 2). The protective antigen (PA), one of the three toxin proteins, induces neutralizing antibodies. The PA in itself is not toxic, but attaches to a receptor on the host cell. After proteolytic cleavage by the host enzyme furin, PA forms a membrane-inserting heptamer that translocates the toxic *B. anthracis* enzymes, oedema factor (OF) and lethal factor (LF), into the cytosol. Combination of the PA with the LF or the OF is necessary for a toxic or lethal reaction.

After internalization, there is an increase in cell permeability to Na^+ and K^+, which is followed by hydrolysis of ATP with production of increased levels of cyclic AMP (cAMP). The resulting cAMP accumulation precedes the later events of calcium influx and cytolysis. The hydrolysis of ATP with the production of cAMP is catalyzed by the OF, a calmodulin dependent adenylate cyclase.

The molecular action of the LF has recently been elucidated [2, 3]. This enzyme is a metalloprotease that cleaves the amino terminus of mitogen-activated protein kinase kinases 1 and 2 (MAPKK1 and MAPKK2). This cleavage inactivates MAPKK1 and inhibits the MAPK signal transduction pathway. This hydrolytic activity is essential for the cytotoxicity of LF.

TABLE 2. Components of Anthrax Toxin

• Protective Antigen
  ▪ not toxic
  ▪ immunogenic
 	
• Oedema Factor
  ▪ causes oedema in combination
   with Protective Antigen
  ▪ limited immunogenic
	• Lethal Factors
	  ▪ lethal in combination with
	   Protective Antigen
	  ▪ limited immunogenic

The crystal structure of PA and its domains have been described by Petosa et al. [4]. The PA is responsible for adhesion of the anthrax toxin to the cell surface and mediates uptake of OF or LF into the cell. These properties of the PA make it a useful candidate antigen for the *Lactobacillus* carrier vaccine approach. With other live vectors, such as *Salmonella*, it has been shown that adhesive capacity is advantageous for eliciting an immune response. In this regard, adhesion may facilitate uptake of the carrier organisms by the M cells in the gut-epithelium. Expression of the PA and subsequent immunization has been carried out previously using gram-negative bacteria as carriers [5]. The gram-positive *Lactobacillus* may be advantageous for expressing *B. anthracis* genes since the codon usage by *Lactobacillus* and *B. anthracis* may be more similar than the codon usage of a gram-negative bacterium.

Existing human vaccines against anthrax show considerable side effects and must be given in multiple booster doses over a period of time, administered parenterally using adjuvants. These vaccines may not be fully protective against all strains of *B. anthracis*, and they are also expensive to produce. Therefore, new generation vaccines are needed against anthrax

1.3 LIVE VACCINES

A new generation of vaccines should meet a number of requirements. Ideally these vaccines should be effective when administered orally in a single dose. They should be safe, cheap, stable and easy to administer. The scientific consensus at this point is that live vaccines are potential candidates for single dose oral immunization. These vectors consist of a relatively non-pathogenic microorganism as a carrier of foreign genetic material able to direct the synthesis of a protective antigen, which can induce neutralizing immune responses against selected pathogens.

Such vaccines are also preferable from a safety point of view, as it would no longer be necessary to culture the pathogen itself in large quantities. Culturing of dangerous pathogens requires stringent containment facilities and strict working protocols for the people involved in the production and downstream processes. A carrier vaccine will thus be much cheaper, safer with regard to production and also with respect to the induction of unwanted side effects. An additional advantage is the fact that countries who want to develop a covert offensive biological weapon program will no longer be able to claim that they are pursuing large-scale production of pathogens for vaccine purposes.

A major scientific advantage of using live microorganisms administered orally is the fact that these vaccines are delivered to and induce immunity at the mucosal surface, the site where most infections actually occur and where the first line of defense against infection lies. Although some success has been obtained with oral vaccines, a disadvantage until now has been the fact that the carriers used (e.g. *Salmonella*, *Escherichia coli*, Vaccinia virus) cannot be classified as safe. Furthermore, these carriers are highly immunogenic by themselves, thereby causing unnecessary activation of the immune system against the carrier and possibly hampering repetitive use of the vaccine carrier with other antigens.

Oral vaccines may stimulate the mucosal immune system and induce a better protection against pathogens that attack through the mucosa. *Lactobacillus* used as a vector for expressing foreign proteins from pathogens may provide such a new generation of vaccines.

Oral administration is a most convenient method of vaccination as compared to the more frequently used parenteral routes, mainly because of the less stringent criteria for a) application (feeding vs. injection), b) education of trained healthcare workers and c) the introduction of both mucosal and systemic immune responses. For the above and other reasons it would be preferable to use a GRAS (Generally Regarded As Safe) organism with probiotic properties, e.g. *Lactobacillus*, as a safe vector for the delivery of foreign antigens to the gastrointestinal tracts of animals and humans (Table 3).

TABLE 3. Advantages of *Lactobacillus* as Vaccine Carrier

- GRAS status
- Natural commensal of the gut
- Probiotic properties
- Adjuvant properties
- Safe production
- Inexpensive production
- Stable
- Easy to administer
- Multi-carrier approach feasible

1.4 THE MUCOSAL IMMUNE SYSTEM

The mucosal surfaces of the gastro-intestinal, respiratory and urogenital tracts have a combined surface area of at least 400 m^2, covered largely by a single layer of epithelial cells. These cellular barriers face environments rich in pathogens, and for defense reasons mucosal tissues are heavily populated with cells of the immune system. The intestinal lining contains more lymphoid cells and produces more antibodies than any other organ in the body. In general, the mucosal immune system represents the first line of defense against pathogens attempting to enter the body. The exposure to an antigen at one mucosal site leads to immunological protection after renewed exposure to the same antigen, even when presentation of that antigen occurs at a distant mucosal surface. In this regard, all mucosal surfaces form an integral, communicative, specialized lymphoid system. Oral vaccination and presentation of the antigen to the intestinal mucosa will also generate protective immunity at the mucosa of the salivary and lacrimal gland, the respiratory tract and the reproductive system. In addition, systemic IgG and T-cell responses are induced.

Intestinal antigen uptake primarily occurs through microfold M-cells which are specially adapted for IgA mediated transport from the lumen to the intestinal lymphoid cells. Following antigen processing across the epithelial barrier and presentation in inductive sites, antigen specific B-lymphoblasts which are IgA committed proliferate locally and then migrate via the regional lymph nodes, thoracic duct and blood stream to local and distal mucosal sites. They differentiate primarily into polymeric IgA producing plasma cells. Dimeric or polymeric IgA antibodies are transported across epithelial cells into glandular and mucosal secretions via receptor mediated transcytosis.

Because most infections begin at mucosal sites, an attractive strategy for preventing such infections is to induce a protective immune response to pathogens when they contact mucosal surfaces. The methods to realize this are still being developed.

1.5 *LACTOBACILLUS* AS A LIVE VACCINE CARRIER

Lactobacillus is a gram-positive, rod-shaped, non-pathogenic microorganism, which belongs to the lactic acid bacteria. This group of bacteria has been used for a long time in the preparation and processing of food and beverages, and the organisms also have health-promoting properties. The use of *Lactobacillus* as a live vector offers some major advantages over other microorganisms. 1) *Lactobacillus* is a GRAS organism, 2) it possesses intrinsic adjuvanticity, 3) it is a normal commensal of the gut, 4) the antibody response against *Lactobacillus* itself is low and 5) *Lactobacillus* mediates numerous additional probiotic effects. In our view these characteristics make *Lactobacillus* an ideal vaccine carrier for oral administration (Figure 1).

Figure 1. The oral vaccine as envisioned.

The application of *Lactobacillus* as an oral vector is, obviously, determined by a complex set of factors, one of which is the capacity of a selected *Lactobacillus* strain to efficiently produce the antigen of choice. Antigen presentation, i.e., how the antigen should be targeted to the immune system, is important. Therefore, different vectors have been developed for intracellular expression of the antigen, surface-associated production of the antigen or secretion of the antigen [6].

Several strains are available for transformation, each differing somewhat in its properties. At the moment, it is not clear whether colonization is a prerequisite for a protective immune response or whether colonization may induce tolerance. Strains with a high adjuvant capacity are to be preferred, and the target animal also determines the choice of the particular *Lactobacillus* strain.

In the following we describe the expression of PA by *Lactobacillus casei*, the first immunization experiments with this model carrier vaccine and an antibody response evaluation.

2. Results

2.1 EXPRESSION OF THE PROTECTIVE ANTIGEN BY *L. CASEI* TRANSFORMANT B-PA3.1

The *pag* gene encoding the Protective Antigen (PA) from *Bacillus anthracis* was amplified by the polymerase chain reaction (PCR) from pPA101 [7] using primers extended with appropriate restriction sites suitable for further cloning. The *pag* gene was cloned into the plasmid pLP503t [6], a plasmid with endogenous regulation elements from *Lactobacillus* used for intracellular expression. The constructs were characterized by evaluation of restriction patterns. The newly formed plasmid, B-PA3 (Figure 2) was transformed to *L. casei 393*.

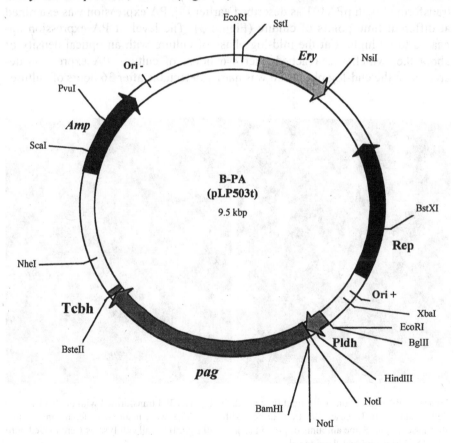

Figure 2. Schematic representation of pB-PA3.1 constructed for use in Lactobacillus.
pag: gene encoding the Protective Antigen from Bacillus anthracis; Pldh: promotor of the lactate dehydrogenase gene of L. casei; Tcbh: terminator of the conjugated bile acid hydrolase of L. plantarum

Transformants were selected on the basis of PA expression detected by Western blotting analysis. Expression of PA was measured in cell lysates from *Lactobacillus casei* transformant B-PA3.1. The proteins of a cell lysate were separated in sodium dodecyl sulfate polyacrylamide gel electrophoresis (SDS-PAGE, 10-15% gels) using the Phast System (Pharmacia, Uppsala, Sweden) and subsequently blotted onto a nitrocellulose membrane. The PA protein in the blots was detected with a mouse monoclonal antibody (MAb C3) specific for PA, provided by Dr. Baillie, Porton Down, UK. Bound antibodies were detected with goat-anti-mouse IgG conjugated to alkaline phosphatase. Staining was performed using BCIP/NBT as substrate. SDS-PAGE gels were silver-stained using the protocol of the Phast System.

A band of the correct size for PA was seen on the blot compared to the recombinant protein (recPA). RecPA was purified from *Bacillus subtilis* strain transformed with pPA101 as described earlier [8]. PA expression was examined at different time points of culture (Figure 3). The level of PA expression appeared to be highest at the mid-log phase of culture with an optical density of about 0.6, which was reached after 4-6 hours of culture. PA expression decreased at the end-log phase and was hardly detectable after 36 hours of culture.

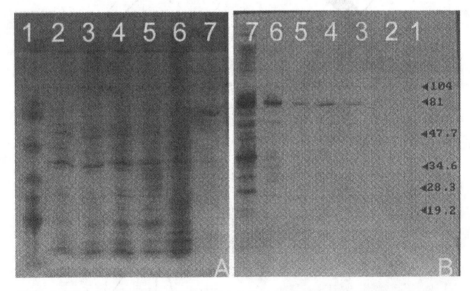

Figure 3. PA expression at different time points by L. casei 393 transformed with pB-PA3.1.
Cell lysates from L. casei 393 transformed with pB-PA3.1 were made at different time points after inoculation. Same amounts of protein (1 μl of 30 μg/ml) for all cell lysates were loaded onto a 10-15 % SDS-polyacrylamide gel.
A) SDS-PAGE. Protein bands were made visible by silver staining. B). Western blot. The PA band was detected using a mouse monoclonal antibody specific for PA. Lanes: 1) pre-stained marker, 2) at 32 hours of culture, 3) at 16 hours, 4) at 12 hours, 5) at 8 hours, 6) at 4 hours, 7) recPA.

2.2 IMMUNIZATION OF MICE WITH A CELL LYSATE OF *L. CASEI* TRANSFORMED WITH B-PA3.1

To investigate whether the PA expressed by *Lactobacillus* could induce an antibody response specific for the PA protein, a group of 3 female BALB/c mice (12 weeks old) was immunized intraperitoneally at day 0 and 28 with a cell lysate (100 µg protein) prepared from the *L. casei* transformant. The cell lysate was admixed with specol as adjuvant [9], and a total volume of 250 µl was administered. PA-specific antibody responses (IgG) were determined by ELISA at day 14, 28, 35 and 42 after the first immunization. PA-specific antibody responses were detectable for the first time on day 28. The PA-specific antibody level increased considerably after a booster immunization (Figure 4, see next page). The results showed that PA expressed by *L. casei* could induce PA-specific antibody responses.

For measuring antibody responses in ELISA, PVC microtiter plates were coated overnight at 4°C with recPA (5 µg/ml). The plates were blocked with 0.5% gelatin in PBS. Sera from mice were prepared inlog to the base 2 serial dilutions starting with a 1:25 dilution of serum in PBS containing 0.1 % gelatin and 0.05 % tween-20 and incubated for one hour at 37°C with the recPA coated on the microtiter plates at 37°C. Bound antibodies were detected using goat-anti-mouse IgG or Ig conjugated to alkaline phosphatase (KPL, Kirkegaard, Denmark) with an incubation period of one hour at 37°C. P-nitrophenolphosphate (1 mg/ml in DEA/MgCl$_2$, pH 9.8) was used as substrate. The absorbance was read after 30 minutes at 405 nm. Normal mouse serum was used as negative control (background). Background signals were subtracted from serum signals.

2.3 IMMUNIZATION OF MICE WITH LIVE *L. CASEI* TRANSFORMED WITH B-PA3.1

In order to obtain mucosal immune responses, mice were immunized orally or intranasally at days 0, 1 and 2 with 5 x 10^{10} or 1 x 10^9 (oral), or 4 x 10^9 or 2 x 10^7 (nasal) live *L. casei* transformed with B-PA3.1, or with the wild type *L. casei*. A booster series was given at days 28, 29 and 30. PA-specific antibody responses could not be detected at days 14, 28, 35 and 42 after the first immunization. *Lactobacillus*-specific antibody responses could be detected after the booster immunization in some mice. The *Lactobacillus*-specific antibody responses in sera from mice immunized orally or intranasally were considerably lower than the antibody responses in sera from mice immunized intraperitoneally with a bacterial lysate (data not shown).

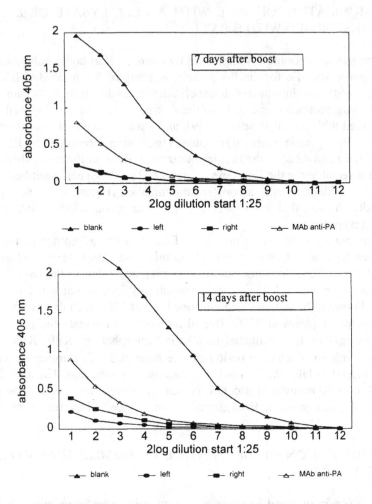

Figure 4. PA-specific antibody response in sera from mice immunized with cell lysate.
The PA-specific antibody response in sera from 3 mice immunized with a cell lysate of L. casei 393 transformed with pB-PA3.1 is shown. Log to the base 2 serial dilutions of sera were made starting at a dilution of 1:25. The upper graph shows the antibody response 7 days after a booster immunization, the lower graph shows the antibody response 14 days after a booster immunization.

3. Discussion

The Protective Antigen (PA) from *Bacillus anthracis* was cloned into a plasmid suitable for *Lactobacillus,* and the construct was subsequently transformed to *Lactobacillus casei.* We showed that *L. casei* is able to express the PA of *B.*

anthracis; the level of expression is high at the mid-log phase and decreases at the end-log phase of culture. It is not yet clear whether this is due to protein breakdown by proteases produced in the later phase of culture. A cell lysate from *L. casei* expressing the PA was able to induce a PA-specific antibody response in mice immunized intraperitoneally. These results indicate that i) the PA molecule expressed by *L.casei* is similar to the PA molecule expressed by other bacteria, and ii) PA contained in a mixture of *Lactobacillus* proteins is immunogenic, and antibody responses in mice are not dominated by *Lactobacillus* proteins.

We did not detect PA-specific antibody responses after oral and nasal immunizations with *L casei* transformed with B-PA3.1. We did detect *Lactobacillus*-specific antibodies in sera after oral and nasal immunizations indicating antigen uptake, processing and presentation by the mucosal immune system. Lack of PA-specific responses may be caused by i) immunodominant proteins from *L. casei*, or ii) an insufficient level of PA expression at the time of antigen uptake and processing.

Since *L. casei* readily accepts ligated DNA material, it is the first *Lactobacillus* species of choice to transform. Plasmids propagated in *L. casei* have recently been purified from *this organism* and subsequently transformed to other *Lactobacillus* species. Immunization experiments with these new transformed *Lactobacillus* species have been scheduled. Other *Lactobacillus* species differ in properties such as adjuvanticity, protease activity and persistence in the gut [6, 10], and these properties may well contribute to vaccine suitability. From our experiments with the model antigen TTFC we have seen that a *Lactobacillus* species with an elongated persistence in the gut and a protease activity different from that of *L. casei* was capable of inducing a TTFC-specific antibody response after mucosal immunization. By contrast, a *L. casei* transformant failed to induce a TTFC-specific response after oral immunization (Shaw, personal communication).

In this chapter we described the first steps in the development of a live oral vaccine based on *Lactobacillus* as a carrier/expression system. Many questions relevant to the production of a good mucosal vaccine using *Lactobacillus* as a vector for expressing pathogenic antigens still have to be investigated in the near future. Parameters such as dosage of microorganisms, level of PA expression, optimal time of administration i.e. stage of cultured Lactobacilli, dead versus live bacteria, the *Lactobacillus* strain, the way of antigen presentation by *Lactobacillus,* and immunomodulation by e.g. cytokines will be studied. In addition to evaluation of PA-specific antibody responses in sera, we will also investigate responses in other samples such as saliva, feces, lung lavages, lymphatic tissues, and other immunocorrelates, such as cell-mediated immune responses. Subsequent to evaluation of immune responses, we will perform protection studies in guinea pigs.

196

4. References

1. Turnbull, P.C.B. (1990) Anthrax, in M.T. Parker and L.H. Collier, (eds.), *Topley & Wilson's Principles of Bacteriology, Virology and Immunity*, 8th edition, volume 3, Edward Arnold, London, Melbourne, Auckland, pp. 366-378.
2. Duesbery, N.S., Webb, C.P., Leppla, S.H., Gordon, V.M., Klimpel, K.R., Copeland, T.D., Ahn, N.G., Oskarsson, M.K., Fukasawa, K., Paull, K.D., and Vande Woude, G.F. (1998) Proteolytic inactivation of MAP-kinase-kinase by anthrax lethal factor, *Science* **280**, 734-737.
3. Vitale, G., Pellizzari, R., Recchi, C., Napolitano, G., Mock, M., and Montecucco, C. (1998) Anthrax lethal factor cleaves the N-terminus of MAPKKs and induces tyrosine/threonine phosphorylation of MAPKs in cultured macrophages, *Biochem. Biophys. Res. Commun.* **248**, 706-711.
4. Petosa, C., Collier, R.J., Klimpel, K.R., Leppla, S.H., and Liddington, R.C. (1997) Crystal structure of the anthrax toxin protective antigen, *Nature* **385**, 833-838.
5. Coulson, N.M. (1994) *Bacillus anthracis* protective antigen, expressed in *Salmonella typhimurium* SL 3261, affords protection against anthrax spore challenge, *Vaccine* **12(15)**, 1395-1401.
6. Pouwels, P.H., Leer, R.J., and Boersma, W.J.A. (1996) The potential of *Lactobacillus* as a carrier for oral immunization: Development and preliminary characterization of vector systems for targeted delivery of antigens, *J. Biotechnol.* **44**, 183-192.
7. Ivins, B.E., Welkos, S.L., Knudson, G.B., and Little, S.F. (1990) Immunization against anthrax with aromatic compound-dependent (Aro-) mutants of *Bacillus anthracis* and with recombinant strains of *Bacillus subtilis* that produce anthrax protective antigen, *Infect. Immun.* **58**, 303-308.
8. Miller, J., McBride, B.W., Manchee, R.J., Moore, P., and Baillie, L.W. (1998) Production and purification of recombinant protective antigen and protective efficacy against *Bacillus anthracis*, *Lett. Appl. Microbiol.* **26**, 56-60.
9. Bokhout, B.A., Van Gaalen, C., and van der Heijden, Ph.J. (1981) A selected water-in-oil emulsion: Composition and usefulness as an immunological adjuvant, *Vet. Immunol. Immunopathol.* **2**, 491-500.
10. Claassen, E., Pouwels, P.H., Posno, M., and Boersma, W.J.A. (1994) Development of safe oral vaccines based on *Lactobacillus* as a vector, in E. Kurstak (ed.), *Recombinant Vaccines: New Vaccinology,* Int. Comp. Virology Org., Montreal.

VACCINE PRODUCTION

JACK MELLING
The Salk Institute for Biological Studies
Biologicals Development Center
PO Box 250, Swiftwater, PA 18370, USA

1. Introduction

The biotechnology revolution which began in the 1970s with the advent of genetic engineering, coupled with major developments in fermentation and downstream processing technology, has had major impacts on research, development
and production of vaccines. In the case of vaccine production, the effect of recombinant DNA techniques has been to make vaccines available against agents
where previously production and isolation of key antigens in sufficient quantities was either not possible or very difficult and expensive. The now widespread
use of hepatitis A and hepatitis B vaccines has been made possible by transferring genes coding for viral antigens to bacteria or yeast, which can then be cultured, and the antigens purified much more effectively than would be possible
with the original viruses. In addition, advances in production methodologies for
more traditional vaccines have resulted in improved yields, ease of manufacture
and better adherence to current Good Manufacturing Practice (cGMP) [1].

Some examples of products which can be manufactured as a result of this
new biotechnological capability are shown in Table 1.

Conversely, the ability to move genetic material from one microorganism to
another has meant that it is possible to create organisms which have multiple
antibiotic resistance, are refractory to vaccine prophylaxis, or contain additional
pathogenic determinants. Furthermore, as our understanding of the human immune system's response to infection develops, it is likely that vaccines can be
designed specifically to enhance the appropriate immune responses. Again there
is the caveat that an aggressor could use such knowledge to engineer organisms
which would be refractory to vaccine protection.

A. Kelle et al. (eds.), The Role of Biotechnology in Countering BTW Agents, 197–214.
© 2001 *Kluwer Academic Publishers. Printed in the Netherlands.*

TABLE 1. Examples of products of current
biotechnological methodologies

Human Insulin	Erythropoietin
Factor VIII	Interferons
Human Growth Hormone	Hepatitis A Vaccine
Tissue Plasminogen Activator	Calcitonin
Gonadotrophins	Human Albumin
Interleukins	Soluble CD4
Hepatitis B Vaccine	Antibodies (IgG)

In view of the number of potential Biological Warfare (BW) agents which already exist (see below), further expansion by producing recombinant strains which could be resistant to immune responses elicited by current vaccines would re-emphasize the need for a fundamental re-think on the role of vaccines for protection against BW. The regulatory requirements which have to be met in order to have product approval for and manufacture of such vaccines mean that the pace of vaccine development may not be able to match the weaponization ability of an aggressor. The problem is difficult enough if it is only natural, unmodified agents which are involved, but would be compounded if organisms refractory to vaccine protection can be produced.

It is salutary to consider that for the potential threat agents on the US DOD list (Table 3) only three licensed vaccines have been produced in over 30 years.

Efforts are underway to understand and identify generic mechanisms by which organisms cause disease. Although this work is at an early stage, it may provide the best hope for generic immuno-prophylaxis and therapy to keep pace with the ability of potential aggressors to enhance the pathogenicity of microbes.

2.   Potential Biological Agents

There are many agents that have been considered as potential biological and toxin weapons against humans. A total of 28 is listed in Table 2. Of these there are few for which fully approved vaccines are currently available. Even in many of these cases the ability of some vaccines to protect against exposure by the respiratory route is uncertain. It is clear therefore that to research, develop and manufacture fully approved vaccines against all of these potential agents, including protection against inhalation, will be an enormous task. If the possibility of genetic modification of known strains also needs to be taken into account [2, 3] then a vaccine program to achieve all of the goals becomes almost impossibly ambitious.

TABLE 2. Some potential biological agents against man

Agent	Aerosol Infective Dose (if known)	Incubation Period (Natural) (days)	Vaccine Protection (Known strains)	Effectiveness of Sero- or chemo-therapy (Known strains)
A. Viruses				
1. Yellow fever		3-6	+	−	−
2. TBE		7-14	+	−	−
3. JE		5-15	+	−	−
4. Dengue		5-7	−	−	−
5. VEE	10-100	2-8	+[b]	−[a]	−
6. EEE	10-100	7-14	+[b]	−[a]	−
7. WEE	10-100	7-14	+[b]	−[a]	−
8. Chikungunya		3-12	+[b]	−[a]	−
9. O'nyong-nyong		3-12	+[b]	−[a]	−
10. Rift Valley		3-12	+[b]	−	−
11. Influenza		1-3	+	−	±
12. Hanta		14-28	−	−	−
13. Smallpox	10-100	7-17	+	±	±
14. Crimean-congo	1-10	4-21	−	±	±
15. Lassa	1-10	4-21	±	±	−
16. Ebola	1-10	4-21	−	±	−
17. Marburg	1-10	4-21	−	±	−
B. Bacterial				
1. Plague	100-500	5-60	±	−	+
2. Anthrax	8,000-50,000	1-5	+	−	±
3. Tularemia	10-50	2-10	+	−	+
4. Brucellosis	10-100	5-60	−	−	+
5. Typhoid		3-90	+	−	+
6. Botulism		1-5	+	±	−
7. Staph enterotoxin		1-6 hours	+[c]	+[c]	−
C. Rickettsial				
1. Typhus		7-14	−	−	+
2. Rocky Mountain		3-14	−	−	+
3. Q fever	1-10	10-40	+[b]	−	+
D. Fungal				
1. Coccidioido-mycosis		7-28	−	−	+

Notes:

[a] passive protection for accidental laboratory exposure
[b] vaccine is experimental—used in humans
[c] only tested in animals
[d] + = suitable/good: ± = moderately effective: − = unsuitable or poor

A more limited program to develop, produce and license vaccines against agents of most concern (Table 3) was initiated in 1997 by the US Department of Defense, and called the Joint Vaccine Acquisition Program (JVAP). The aim of the JVAP is, over a 10-year period, to create stockpiles of licensed vaccines for use by the military. Even this program, which addresses vaccines for which much of the development work has been done, is likely to cost several hundreds of millions of dollars. In over 30 years licensed vaccines have only been produced against three of these agents, although all of the agents in Table 3 were covered in the USA's former offensive program.

TABLE 3. BW agents and diseases of concern

Anthrax	Botulinum toxins
Q-Fever	Ricin toxin
Smallpox	Staphylococcal enterotoxin
Venezuelan Equine Encephalomyelitis	Western Equine Encephalitis
Eastern Equine Encephalitis	Brucellosis
Plague	Tularemia

Data derived from the US Department of the Army, Joint Program Office for Biological Defense Request for Proposal DAMD17-95-R-5020 for manufacture of defense vaccines

A further complication is the potential ability of an aggressor to modify microorganisms to make them resistant to antibiotics and/or refractory to vaccine protection. It has been reported that the former Soviet Union carried out such developments in respect of plague, anthrax, tularemia, glanders and smallpox [3, 4].

In addition, the major advances in fermentation technology during the last 20 years have substantially improved the capability to culture bacteria in high yields. Perhaps more significantly, the progress which has been made in animal cell culture techniques means that many viruses can now be produced on a scale which would have been hard to imagine when the Biological and Toxin Weapons Convention (BTWC) was signed. Such improvements in cultivation techniques, coupled with the ability to create chimaeric organisms, could significantly expand the number of agents of concern.

3.   Types of Vaccines

The aim in developing and producing a vaccine is to deliver an antigen (or antigens) so as to stimulate an immune response which will protect against the relevant organism [5]. Over the years a number of different types of vaccines have been produced with newer products, and these approach the ideal of delivering only those antigens specifically required for protection. Nevertheless, the history of BW vaccine development means that a future vaccine portfolio is likely to encompass vaccines of many types, thus adding to the complexity of production and also of vaccine administration.

3.1 WHOLE KILLED VACCINES

These consist of the whole organism which has been killed by heat or chemical (usually formalin) treatment. This category of vaccines includes plague, the older typhoid and cholera vaccines, and eastern and western encephalitis. Generally such vaccines, particularly those of bacterial composition, have been more reactogenic and less effective in stimulating specific protective immune responses.

3.2 SUBUNIT VACCINES

Instead of the whole microorganism an antigen or collection of antigens is prepared by extraction and purification from either a live or killed organism. Subunit vaccines tend to be less reactogenic and produce a more specific immune response. Inclusion of an adjuvant (usually aluminum hydroxide) is normally required.

Antigens for subunit vaccines, which also include the traditional toxoid vaccines (diphtheria and tetanus), may be prepared by extraction and purification directly from the microorganism or by cloning the relevant gene(s) into another organism. In the case of hepatitis A and B vaccines the viral surface antigens are expressed in yeast and then purified.

3.3 SYNTHETIC PEPTIDES

It is possible to identify regions of the primary sequence of protein antigens which form epitopes against which neutralizing antibody can be formed. Such peptides, some as little as 20 amino acids or less in length, can be chemically synthesized, and when linked to carriers, as first demonstrated for foot and mouth disease virus, can produce immunity to challenge. The present limitations are the need for carriers and the fact that some epitopes are made up of parts of the primary sequence which are not contiguous, but only come together in the tertiary structure and are not amenable to synthesis.

3.4 LIVE ATTENUATED VACCINES

Some of the most successful vaccines consist of living organisms in which the virulence is attenuated. These vaccines stimulate both humoral and cell-mediated immune responses which closely resemble those induced by natural infection. Attenuated strains have been found naturally occurring (vaccinia), selected by animal or culture passage (BCG, measles, yellow fever, rubella), or specifically attenuated by deletion of some key portion of the viral genome (pseudorabies) [6].

Recombinant DNA technology has also enabled the basis of attenuation of the older vaccines to be better understood and the risk of reversion to be assessed. This technology has also enabled a new class of live attenuated organisms to be developed by inserting genes from other organisms into some vector. One of the most studied vectors is vaccinia virus, and recombinant vaccinia viruses have been derived which express immunogens for, among others, influenza, rabies, and hepatitis B viruses. Other potential vectors include herpes simplex, various alphaviruses and avian pox viruses, although no product using such vectors has yet gained regulatory approval.

3.5 DNA VACCINES

DNA vaccines represent a rapidly developing technology in which protein antigens can be expressed in vivo from plasmid DNA taken up by animal cells [7]. Animal model experiments have shown that the expressed antigens can generate strong and long-lasting immune responses. In addition, DNA vaccines have been shown to induce both humoral and cell-mediated immune responses.

4. Regulatory Compliance for Manufacture

In order for vaccines, as for other medicinal products, to be made generally available they must meet the criteria of quality, safety and efficacy required by regulatory agencies. This is equally true for defense vaccines, since governments in the USA and European countries have decided not to make any large-scale use of unlicensed products for protection either of civil or of military populations. Given the potentially large number of vaccines involved and their current state of development, achieving product approval and production in accord with current cGMP is a considerable challenge [8].

cGMP embodies "a set of scientifically sound methods, practices or principles that are implemented and documented during development and production to ensure consistent manufacture of safe, pure and potent products." Such principles apply to the manufacturing process as well as to facilities and equipment in which products are manufactured.

The principles on which cGMP is based are that (i) quality, safety and effectiveness must be designed and built into the product; (ii) quality cannot subsequently be inspected or tested into the product; (iii) each step in the manufacturing process must be controlled to ensure that the final product meets quality and design specifications; (iv) there is documentary evidence showing compliance with cGMP.

Accordingly, vaccine production is very highly regulated so as to ensure that the products are of consistent quality, safe, and effective for the purpose(s) for which regulatory approval was granted. Therefore the cost of maintaining facilities, including the necessary quality and regulatory operations, is considerable. In the case of vaccines for civil use involving millions of doses of any one vaccine, the cost can be spread across a wide product base. In contrast, the required volumes of any one defense vaccine are likely to be one tenth to one hundredth of those for civil vaccines. In addition, there is a wide range of techniques involved in producing viral, bacterial, rickettsial and fungal vaccines coupled with the need for levels of biocontainment which exceed those normally encountered in the vaccine industry. Taken together these factors present unique problems in the manufacture of defense vaccines.

There are several key aspects which must be considered in respect of regulatory compliance. These are the facilities, including buildings and equipment; operating procedures used in the manufacturing process; and the products themselves.

4.1 FACILITIES

The essential principle in the design of a facility for production of biological products is to minimize the potential for mix up of one product with another or contamination of a finished or late stage product with starting materials. Thus segregation, both physical and temporal, of different products and stages of manufacture is vital. Physical segregation is achieved by designing the facility so that product flow patterns ensure there is no cross over between early and later stages of processing. In addition the building design should address the movement of personnel, raw materials, equipment and waste materials. Normally these flow control concepts are part of the building design. Isolation of different manufacturing stages is further aided by having air handling equipment which is dedicated to specific manufacturing areas such as fermentation, downstream processing, filling and lyophilization. Air pressure differentials between the various processing areas also contribute to effective segregation. Clearly the quality of the air supplied to process areas is critical in minimizing opportunities for either microbial or particulate contamination.

Cleaning is of major importance and surfaces of walls and floors should be non-porous so as to be resistant to frequent exposure to materials used for cleaning and/or decontamination. To aid cleaning, surfaces should not have ledges or cracks, and pipework or other services, which are hard to clean, should be enclosed to avoid a build up of particulates.

There are stringent criteria for the quality of steam and water used in bioprocessing which may go into the product or have contact with equipment used in the production process. There is therefore a requirement for pure steam generators that are designed for the production of Water for Injection (WFI) quality pyrogen free steam for use in pharmaceutical and biotechnological processes.

Plant steam is unacceptable for use in the previously mentioned applications. This is due to the presence of contaminants including boiler treatment additives along with the potential for pyrogenic contamination. A pure steam generator can use the plant steam as a primary energy source because the plant steam never comes into contact with the pure steam.

Once WFI has been generated its quality, especially freedom from microbial contamination, must be maintained, and normally this is done by installing a continuously circulating loop maintained at 80°C. The system design and operational validation are essential to obtain regulatory approval.

Air quality appropriate to the particular process area is achieved by HEPA filtration of incoming air and also by ensuring, through a number of air changes and pressure differentials, that the particulate count is maintained below the relevant limits.

If live agents are involved in the process, either as the final product in the case of an attenuated organism or as the source of antigens for subunit or killed vaccines, then a further degree of complexity in facility design will be involved. Not only must the facility meet cGMP standards in respect of the product, it must also be structured to ensure appropriate containment of live organisms. These two criteria are not easily reconcilable, and considerable ingenuity and expense will be involved in designing and building facilities to meet the two requirements.

4.2 EQUIPMENT

The issues relating to equipment are essentially the same as for buildings in that the aim is to prevent cross contamination or confusion of one product with another. In the case of measuring equipment to ensure that the readings are correct, such equipment must be validated at appropriate intervals.

4.2.1 Shared Equipment

This is equipment which is used with different products and/or at different process stages with the same product. The essential requirement is to have in place validated cleaning or decontamination procedures and to have Standard Operating Procedures (SOPs) for the changeover operations. Typically items such as fermenters, centrifuges, lyophilizers and filling machines fall into the category of shared equipment. Equipment or buildings which have been used for work with spore forming organisms cannot subsequently be used for pharmaceutical production of products from non-sporeformers.

A major difficulty to be overcome with any equipment previously used in making one product will be clearly to demonstrate that microorganisms or toxins to which it was exposed have been inactivated and removed. In view of the likely agents involved there will be great concern lest any subsequent products become contaminated. If clean up and decontamination cannot be validated, such equipment will be of no use for campaign production.

4.2.2 Dedicated Equipment

This may only be used for one process stage of one product, although it can be re-used for the same process stage of subsequent batches of the same product. Obviously cleaning is required, but the requirements are less onerous than for-shared equipment.

4.2.3 Disposable Equipment

Such equipment is discarded after use and mainly consists of pipettes, syringes and various plastic or glass items.

As with facilities some equipment will need to be specifically designed to achieve containment of hazardous microorganisms or toxins while also meeting cGMP requirements. One example is the design and construction of a 150 liter fermenter for culture of organisms requiring BSL-3 containment [9], and a number of other examples have also been described [10].

4.3 OPERATING PROCEDURES

In a pharmaceutical facility every activity which can impact on product quality needs to be the subject of an SOP. As has been mentioned above, this requirement goes well beyond the production process itself and covers most aspects of the facility's operation.

The activities covered by SOPs include raw materials testing and storage, equipment and facility operation, production methodologies, and all aspects of quality assurance and quality control.

4.4 VALIDATION

An essential facet of the cGMP process is that of ensuring i.e. validating that all elements involved operate according to specification. Validation comprises a defined strategy of inter-related practices and procedures which in combination with routine production methods and quality control techniques provides documented assurance that a system is performing as intended, and/or that a product conforms to its pre-determined specifications. Overall there are three elements which make up the validation process.

(i) Prospective validation involves establishing documented evidence, prior to process implementation, that a system does what it purports to do based on a preplanned protocol. (ii) Retrospective validation involves establishing documented evidence that a system does what it purports to do based on review and analysis of historical information. (iii) Concurrent validation involves establishing documented evidence that a process does what it purports to do based on information generated during actual implementation of a process.

There are various types of documentation associated with the validation process in order to ensure that a structured approach is observed (Table 4).

TABLE 4. Validation documentation

Validation Master Plan – A technical and administrative resource planning tool for the validation and commissioning process and to provide a guideline for the validation approach for each piece of equipment and process.

Enhanced Turnover Packages (ETOP) – Documentation binders or files which contain all relative design, installation and testing documentation generated during design and construction for a particular engineered system. This document package is intended to support the Installation Qualification. (May be prepared by contractor)

Installation Qualification (IQ) – Documented verification that all key aspects of the installation adhere to appropriate codes, approved design specifications and manufacturer's recommendations (where appropriate).

Operational Qualification (OQ) – Documented verification that systems or subsystems capable of being challenged perform as intended throughout all anticipated operating ranges.

Performance Qualification (PQ) verifies the performance of critical utility systems or processes. Critical utility systems such as WFI and Pure Steam are challenged throughout proposed operating ranges for extended periods of time, while an extensive program of quality monitoring is performed.

Change Control – A formalized program by which qualified representatives review proposed and actual changes to products, processes, equipment or software to determine their potential impact on the validation status.

Acceptance Criteria – A set of measurable qualities or specifications used to check out system or equipment installation, operation or performance. Conformance with these qualities or specifications provides a high degree of confidence that a system is installed, operates or performs as intended.

It is therefore readily apparent that major effort and resources must be devoted to this aspect of the cGMP activity.

4.5 TESTING

The manufacture of any biological product requires close attention. It has long been recognized that, unlike other pharmaceutical products, biological products cannot be tested adequately for quality, safety or efficacy by chemical means alone. More specifically, biological products may be subject to the following difficulties: they may have no chemically defined composition, which could result in undetected and undesirable errors occurring during the production process, and in variation from batch to batch; the only way of qualitatively or quantitatively evaluating biological products is usually by measuring their effect in an in-vivo system; the quality of biological products cannot be guaranteed by a final test result on a random sample of the final product – it may be achieved only by a combination of in-process and end-product tests that monitor the entire production process from beginning to end.

It is for these reasons that a whole battery of traditional safety and quality control tests have been developed and applied over the years to vaccines and other biological products. The ever-increasing concern with safety, fuelled particularly in the western world by the very real fears of litigation when actual or theoretical problems arise, has resulted in the virtual demand for zero-risk products.

The rapid development of biotechnological processes, in particular those involving recombinant DNA technology, hybridoma technology and mammalian cell line technology, has served only to heighten testing requirements. Products from genes expressed in foreign hosts may differ structurally, and therefore differ biologically or immunologically from their natural counterparts; foreign genetic material might find its way into the end-product, and may therefore adversely affect man, animals or the environment; scale-up and unintended variation in culture during production may cause changes that could favor the expression of alternative host-vector genes, alter the product significantly, or increase yields of impurities.

To this end a number of analytical techniques have been or may be brought into play as in-process and end-product tests to address product identity, purity, specificity and stability, the characterization of source (seed) materials and the validation of process procedures.

These techniques complement the more traditional in-process and end-product safety and quality control tests, most of which remain as valid for the new methodologies as for the old.

4.5.1 Seed Stocks

In the case of recombinant DNA host/vector systems, those normally in use currently are bacterial/plasmid, yeast cell/plasmid, mammalian cell/viral and insect cell/viral. All four systems have their advantages and disadvantages, the yeast, insect and mammalian cell systems generally being preferred for eukaryotic gene expression. The recombinant vector itself may be genomic RNA, genomic DNA, messenger RNA, synthetic DNA or any combination of these.

For the host cell itself, the traditional information is still required – origin, source, history, purity, genotype, phenotype, growth characteristics – but in addition for eukaryotic cells other tests are now being applied. These include isoenzyme analysis, DNA fingerprinting and tissue typing antigens.

For the recombinant vector and master and sub-master host-vector cell seed stocks, the range of information and details required is to be found in guidelines issued by the various national regulatory authorities. Of particular importance is the need to monitor the cell seed stocks for vector integrity and stability during storage, recovery and use, preferably for a period of several generations beyond that of the actual production period.

While structural instability is rarely a problem in continuous culture, vector segregational instability, caused by defective partitioning or by a drop in copy number, can be a major potential problem in large-scale fermentation culture. Defective partitioning may not be a problem for a high copy number vector, particularly if careful genetic construction allows inclusion of a partitioning region, but a drop in copy number can be caused by dilution rate, nutrient limitation, increased host cell growth rate, the increased load on the host cell replication system or the toxic effect of the product expressed by the vector.

Vector stability. This is usually monitored by the simple method of screening for associated antibiotic resistance. The percentage of antibiotic resistant cells in the population is related to the number of cells containing the vector. However, it is possible for resistance to be lost without loss of the vector, and neither does the technique give any indication of copy number content.

Copy number. Several methods are available for estimation of copy number, of which the most commonly applied is agarose gel electrophoresis where the vector and chromosomal DNA are separated by their respective molecular weights. The gel is stained, photographed, scanned by densitometry and, with the use of internal standards, accurate measurements of copy number can be calculated.

Integrated vector. Where the vector is in an integrated rather than extrachromosomal state, identification and characterization of sequences is more difficult. This may be achieved by Southern blot analysis of total cellular DNA, Northern blot analysis of product-related transcription, or most rapidly and efficiently currently by polymerase chain reaction amplification techniques [11].

4.5.2 Fermentation

Ever-closer control over fermentation is now accepted as desirable in order to obviate actual or potential alterations in the characteristics of the seed stock or its product(s). As has been noted above, this is particularly relevant in the case of genetically engineered organisms where vector stability and copy number can be affected greatly by dilution rate or nutrient limitations. An essential feature of any monitoring and control systems must be rapidity – a rapid answer allowing a rapid response may be critical for a successful fermentation.

Techniques in use include:

Immunoassay. Current technology allows sensitivities at femtomolar levels and results that may be available within a few minutes. Immunoassays, for example ELISA or Western blot, are particularly valuable for analyzing fermentation broths, when the product to be measured may be low in concentration in the presence of excess of other compounds. This is also true for harvesting and the initial stages of purification procedures [14]. Since such assays are quantifiable they may be used to measure specific production rates throughout the manufacturing cycle; hence they may indicate accurately any fall-off due to unforeseen changes in the seed source or environmental conditions.

Off-gas analysis. This is increasingly widely used to monitor and control fermentation processes automatically. Off-gas composition may give exact indications of the condition of a culture and hence the need for appropriate actions to be taken. The analyses are performed by mass spectrometry linked to computer-activated fermenter control systems.

Biosensors. A wide range of enzyme-sensitized detector systems with desirable selectivities has been developed at research laboratory level, but few have yet found application in fermentation monitoring and control, primarily due to low reliability characteristics, sterilization/sanitization difficulties, fouling problems and technical instrumentation drawbacks. However, it is to be expected that these technical difficulties will be overcome and biosensors capable of monitoring such criteria as biomass density, carbon source availability, dissolved gases, product, staling compounds, cations, trace elements etc., can be envisaged as playing an important future role in quality control monitoring of fermentation and downstream processing stages in production [12].

4.5.3 Downstream Processing

Unlike small molecular weight drugs, biopolymer products embrace materials that may range from simple peptides through high molecular weight glycoproteins to complex formulations of microbial components. The protein molecules may contain disulfide bonds, carbohydrate side chains or acetylated *N*-terminal amino groups and be subject to post-translational alterations. Various methods are available for assessing the identity and purity of these biopolymers [13,14], be they of natural or recombinant origin. In the latter instance there are very specific problems that can occur, such as: proteolytic cleaving of *N*- or *C*-terminus sites, incorrect glycosylation or other post-translational modifications, incorrect disulfide bond formation, amino acid substitution, oxidation, deamidation, incorrect folding after solubilization, or aggregation.

5. Product Licensing

Vaccines developed or manufactured with a view to BW defense use have not generally been subjected to the full rigor of the current product licensing system. Products which are not currently licensed will have to be treated in the same way as any new product. The most important criteria which such products must meet involve (a) safety tests, including toxicological assessment, initially in animals, but then in man, (b) determination of efficacy, again initially using animal models, followed by human studies, and (c) assessment of product quality, involving data on all aspects of production designed to ensure not only that the product is acceptable according to current scientific standards, but also that the standard of quality can be maintained and reproduced from batch to batch during the life of the product (see above).

5.1 SAFETY

A number of elements are involved in ensuring the safety of vaccines.

Seed lots of the particular organism must be established and subjected to careful scrutiny as to the identity and provenance of the organisms. This is important not only for live attenuated vaccines, but also for killed or subunit vaccines to ensure the presence of appropriate antigens.

The various ingredients used in growth media and materials with which the product comes into contact during processing must be selected with a view to their potential toxicity. This is especially important where carry-over of minute and undetectable amounts into the final product could have adverse consequences. For example, since the recognition of bovine spongiform encephalopathy (BSE), materials of bovine origin from BSE endemic areas are excluded. This has required re-formulation of many culture media for both bacteria and viruses, as well as reappraisal of seed stocks which were produced using bovine products. A further example is the need to eliminate the use of toxic chemicals at the processing stage and to set limits for carry-over of other materials and determine that such limits are well below any toxic levels. Materials originating from living organisms which may be used in the production or purification processes, e.g. animal sera and monoclonal antibodies, must either be demonstrably free from adventitious microorganisms or it must be shown that the procedures used would eliminate the risk of such organisms being carried through into the final product.

Depending upon the type of vaccine, killed, subunit or live attenuated, appropriate tests including animal studies are required to validate the inactivation process or to ensure effective attenuation. In the case of inactivation of organisms or toxoiding of toxins the kinetics of inactivation must be determined to ensure a sufficient margin of safety is built into the process. Because any change in formulation or conditions (e.g. pH, temperature, ionic strength) of inactivation can radically alter the kinetics, great care is needed to ensure reproducibility.

Once a production process has been instituted and animal safety data established, tests in man would be the next stage.

Phase I studies in healthy adult volunteers would be required to show the absence of unacceptable side effects before proceeding to Phase II studies, again with informed consent, in larger numbers of people in the target population for the vaccine. Larger Phase III trials for efficacy would normally be undertaken only after acceptable Phase II studies for safety have been carried out; these studies would also contribute to the data on adverse reactions.

A vaccine developed for BW defense is likely to have been tested only in young healthy adults, and if the target population for civil use were different further studies would be essential. Before carrying out such further studies, it would be necessary to review existing data and ensure its relevance to the current product and to the production process. If such data are not available, or if process changes have occurred, then a complete repeat of animal and human safety tests could be required on newly manufactured material.

5.2 EFFICACY

Once the necessary safety criteria have been established, efficacy of vaccines has normally been determined in Phase III clinical trials. The number of people included, their age, sex and geographical distribution depends on the disease involved and its prevalence in particular populations. Such trials may take a number of years effectively to determine the value of a vaccine.

Where any human experimentation is involved, voluntary, competent, informed and understanding consent is absolutely essential. This was one of the principles set out in the Nuremberg Code by the U.S. judges in 1947 [15]. A further requirement is that no experiment should be conducted where there is an *a priori* reason to believe that death or disabling injury will occur as a result.

In the case of many vaccines against potential BW agents (Table 1), the natural (low) prevalence of the disease and absence of aerosol infection as the natural route is such that no conclusive field trial of efficacy could be carried out.

In most instances it would be unethical and unacceptable to perform human challenge tests and thus the efficacy of many (most) BW vaccines cannot be determined directly for man. In the case of some diseases, field studies could not yield the necessary data on protection. Should any country have such data, it is unlikely to be made available since it would involve admitting the violation of human rights.

Human efficacy data is normally required to licensure. In the absence of this, for the reasons described above, a number of vaccines will not be licensable according to the usual criteria.

Where the precise mode of action of an agent is known and data are available on immunological protection (e.g. botulinum toxin) which allow effective prediction of the level of protection by determining amounts of antibody (IgG), there is a strong case to support licensing on the basis that no human in vivo challenge is needed to confirm the efficacy. In contrast, where more complex immune mechanisms are involved which pertain to other bacterial, viral or rickettsial infections, prediction without Phase III trials presents a much greater challenge in the present state of our knowledge of the human immune system.

The case of type A botulinum toxoid vaccine, which is licensed in the UK, is a good example of the use of immune correlates as surrogate markers to substitute for human epidemiological or challenge data. The identification of appropriate surrogate markers in the case of bacterial, fungal or viral infections is much more complex. However, the value of antisera to prevent or treat a wide range of infectious diseases including those associated with bacteria, viruses and toxins is well established. Many such products are approved for human use in countries with rigorous licensing systems. Accordingly, if a suitable animal model for the disease can be established, the ability of human immune serum from volunteer recipients of trial vaccines to give passive protection, especially against aerosol challenge, would provide a powerful indicator of efficacy.

In the longer term, to increase confidence that products are efficacious based on animal and/or in vitro studies it will be essential to develop a more detailed understanding of the pathogenic mechanisms and determinants of the infectious agent and the host immunological responses to those determinants. On such a basis a detailed and scientifically understandable extrapolation from animal challenge data to predict human efficacy should be achievable. Such studies will need to harness the latest research findings in immunology and will pose both a challenge and an opportunity to make informed decisions on product licensing. Such an approach could also have great value in the civil field, where conducting extensive clinical trials on improved vaccines becomes increasingly difficult as the disease incidence diminishes. Valuable improvements to safety, reactogenicity and efficacy may be impeded if surrogate animal model studies and markets cannot be substituted for epidemiological trials.

5.3 QUALITY

As described above, the detail of the production process is a vital aspect in specifying a biological product. Thus, in determining if a product should be licensed, all aspects of the manufacturing process need to be considered to assess their impact on product quality and reproducibility.

6. Conclusion

Advances in biotechnology have enabled, and will enable, vaccines to be produced against an increasing range of pathogens. However, the potential number of BW threat agents, both traditional and those which may be derived by genetic manipulation, presents a major challenge to the use of vaccines for protection.

Vaccine production, as described above, is a complex and highly regulated process. The response time to produce quantities of a vaccine, even one which has already been fully developed and licensed, can be measured in months. For development and regulatory approval of a new vaccine the timescale is at least an order of magnitude greater. This gives a potential aggressor a significant advantage. Thus for a vaccine strategy to be effective, both stockpiling and immunization of at risk personnel in advance of any attack would be essential.

Given that the number of potential BW agents is large and growing, the logistics of maintaining stockpiles, replacing out-of-date products and operating an immunization program represent a major challenge.

Therefore either the target organisms need to be narrowly defined, which will require good intelligence with the associated risk of an aggressor using an agent against which there is no protection, or a major industry will need to be created for defense vaccine production. Such a program would depend on a major expansion of production capability coupled with the continued willingness of both military and civil personnel to accept numbers of immunizations which may far exceed those carried out for normal protection against community acquired infections.

Consequently it is vital to explore both improved physical protection of at risk personnel as well as other means of mobilizing immune responses by a major effort to understand and apply non-specific immune stimulation.

7. References

1. *Guide to Good Pharmaceutical Manufacturing Practice* (1998) Her Majesty's Stationery Office, London.
2. Weiner, T. (1998) Soviet defector warns of biological weapons, *The New York Times*, February 25th, A1.
3. Alibek, K. (1998) Russia's deadly expertise, *The New York Times*, March 27th, OP-ED.
4. Pomerantsev, A.P., Staritsin, N.A., Mockov, Y.V., and Marinin, L.I. (1997) Expression of cereolysine AB genes in *Bacillus anthracis* vaccine strain ensures protection against hemolytic anthrax infection, *Vaccine* 15, 1846-1850.
5. Robinson, A. and Melling, J. (1993) Envelope structure and the development of new vaccines, *J. Applied Bacteriology Symposium Supplement* 74, 43S-51S.
6. Ellis, R.W. (1988) New technologies for making vaccines, in S.A. Plotkin and E.A. Mortimer (eds.), *Vaccines*, W.B. Saunders Company, Philadelphia, pp. 568-575.
7. Whalen, R.G. and Davis, H.L. (1995) DNA-mediated immunization and the energetic immune response to hepatitis B surface antigen, *Clin. Immunol. Immunopathol.* 75, 1-12.
8. Melling, J. (1997) Defence-related vaccines: product licencing issues and constraints, in B.B.S. Price and R.M. Price (eds.), *Proceedings of the 2nd CB Medical Treatment Symposium*, Applied Science and Analysis, Portland, Maine, pp. 81-85.
9. Hambleton, P., Griffiths, B.J., Cameron, D.R., and Melling, J. (1991) A high containment polymodal pilot plant fermenter – design concepts, *J. Chem. Tech. Biotechnol.* 50, 167-180.
10. Hambleton, P. and Melling, J. (1994) Containment of unit processes, in P. Hambleton, P., Melling, J., and Salisbury, T. (eds.), *Biosafety in Industrial Biotechnology*, Chapman Hall, Glasgow, pp. 129-148.

11. Mullis, K., Faloona, F., Scharf, S., Saiki, R., Horn, G., and Erlich, H. (1986) The polymerase chain reaction, *Cold Spring Harbor Symp. Quant. Biol.* **51**, 263-273.

12. Clarke, D.J., Calder, M.R., Carr, R.J.G., Blake-Coleman, B.C., Moody, S., and Collinge, T.A. (1985) The development and application of biosensing devices for bioreactor monitoring and control, *Biosensors* **1**, 213-320.

13. Geisow, M.J. (1991) Characterising recombinant proteins, *Biotechnology* **9**, 921-924.

14. Smith, B.J. and Perry, M. (1990) Analytical techniques for biotechnology, *Chem. Ind.* **18**, 563-567.

15. Annas, G.J. and Grodin, M.A. (1992) *The Nazi doctors and the Nurenberg Code: human rights in human experimentation*, Oxford University Press, New York.

THE ROLE OF BIOTECHNOLOGY IN PROTECTION AGAINST BTW AGENTS - OVERVIEW FROM A MEDICAL POINT OF VIEW AND IDENTIFICATION OF THE SYMPTOMS

GYÖRGY BERENCSI[1] AND GÁBOR FALUDI[2]
[1]*Division of Virology*
"B. Johan" National Center for Epidemiology
[2]*Medical Research Institute of the Hungarian Defense Forces*
Budapest, Hungary

1. Introduction

Data collected in the past have indicated that biological warfare is useless both strategically and tactically [2, 34, 39]. Recent experiences and the results of scientific research in the field of emerging and reemerging infectious diseases suggest, however, that under special circumstances the misuse of recent medical knowledge or the neglect of basic rules of hygiene and preventive measures [7] may result in the creation of public health hazards. New epidemics of infectious diseases have appeared such as those caused by *Salmonella enteritidis*, the agent of bovine spongiform encephalopathy (BSE) or by filoviruses.

For a clinical virologist who has grown up in a family interested in public health for three generations, close similarities in the problems of re- or emerging infections and those of biological and toxin warfare (BTW) agents can be easily recognized. In some cases accidents involving BTW activities resemble the outbreak of reemerging diseases quite strikingly. Therefore, the activities and duties of personnel with respect to research or countering emerging infectious diseases through preventive measures are similar in both types of events.

Recently a special committee of the U.S. National Science and Technology Council summarized recommendations designed to counter emerging infectious diseases. The recommendations are shown in Table 1, and can be applied if biotechnological and health sciences are misused in terroristic or warfare activities.

New potential dangers threatening public health and the environment could arise through the world-wide use of biotechnology. New agents including genetically manipulated potentially harmful microorganisms, plants and animals can be produced by these technologies. Biotechnology, however, also provides new possibilities for the detection of public health hazards, as in the early rec-

A. Kelle et al. (eds.), The Role of Biotechnology in Countering BTW Agents, 215–225.
© *2001 Kluwer Academic Publishers. Printed in the Netherlands.*

ognition of infectious agents and the detection of traces of BTW agents.

Important new means for prevention and therapy such as vaccination, immunomodulation, or possibilities of post-exposure therapy have also been put at our disposal by the biotechnology revolution. Unfortunately in some cases the potential misuse of these technologies must also be taken into account.

TABLE 1. Recommendations of the Committee on International Science, Engineering and Technology Policy of the National Science and Technology Council [25]

1.	Concerted global and domestic surveillance and diagnosis of disease outbreaks and endemic occurrence. This must entail the installation of sophisticated laboratory capabilities at many centers now lacking them.
2.	Vector management and monitoring and enforcement of safe water and food supplies, and personal hygiene (e.g. Operation Clean Hands).
3.	Public and professional education.
4.	Scientific research on causes of disease, pathogenic mechanisms, bodily defenses, vaccines and antibiotics.
5.	Cultivation of the technical fruits of such research, with the full involvement of the pharmaceutical industry and a public understanding of the regulatory and incentive structures needed to optimize the outcomes.

BTW agents can be considered as "misuse" of the fruits of both scientific research and "sophisticated laboratory technologies". Recent developments of biotechnology in the field involving viruses will be discussed below.

2. New Potential Risks Created by Work Utilizing Contemporary Advances in Biotechnology

As early as 1973 molecular biologists recognized the risks posed by genetically engineered microorganisms, but BTW agents were omitted from the agenda of the Asilomar Conference which dealt with risk issues as well as from early correspondences about such risks, apparently due to difficulties in dealing with an issue that would involve classified data [12].

As mentioned above, the diagnostic virologist must recognize and be able to identify the sources and etiology of emerging infectious diseases [8, 29], and to be aware of the possibility of accidents and potential risks posed by new diseases including those created by terrorists or others intent on the misuse of microbiology, molecular biology and biotechnology.

What kind of risks can be created by research on or production and development of BTW-relevant agents using biotechnology? Some of these are summarized in Table 2.

TABLE 2. Possible BTW-relevant products and uses of biotechnology that may pose risks

---

1. **BTW agent as product :**
   Biologically active recombinant substances such as cyclic oligopeptides composed of D-amino acids;
   Viable recombinant gene vectors (vaccinia virus, adenovirus, retroviruses or togaviruses) expressing biologically active products of the genes e.g. immunomodulators, toxins, regulatory substances or genomes of viable viruses;
   Pathogenic viruses produced or amplified by biotechnological procedures;
   Transgenic plants and animals.

2. **Genetically manipulated organisms for use at different stages of industrial production**
   Amplification or modification of genes coding for a product, including mutations for modifying enzyme activity, fusion with genes of signal proteins for localization of the product;
   Attachment of marker sequences to improve or accelerate purification and concentration of the product.

3. **Technology to apply "in process control" to biotechnology procedures:**
   Improvement of decontamination of wastes by genetically manipulated microroganisms;
   Control of biological activity during processing with genetically manipulated organisms including microorganisms, tissue culture cells, transgene experimental animals, insects or plants;
   Use of molecular reagents composed of DNA molecules which are extremely resistent to decontamination including recombinant DNA probes, products of the polymerase chain reaction (PCR) or ligase chain reaction (LCR), microorganisms produced at the laboratory scale for the production of molecular reagents, e.g. recombinant probes, DNA samples for competitive PCR, and different control DNA preparations.

4. **Chemicals and reagents frequently used in biotechnology:**
   Production of inducers, inhibitors, promoter substances, stains, monoclonal antibodies, phagemids.

---

3. New Diagnostic Techniques for the Detection of Small, Specific - Molecular Differences

New aspects have emerged in the medical and diagnostic industry as a result of the world-wide, routine use of biotechnology. The information collected on enteric viruses detected in food outbreaks [15] is a good example to show the changes in diagnostics and laboratory technology.

Hepatitis A virus, small round-structured viruses, and Norwalk caliciviruses have been identified as causative agents of foodborne diseases in recent years [15]. The small, round-structured viruses and the Norwalk caliciviruses cannot be cultivated. Biotechnology provides a means to detect the etiology (cause) of the outbreaks and verify their presence in the food specimen without the necessity of cultivating the organisms [11].

The majority of epidemiological data on non-cultivable viruses causing enterocolitis was collected in recent years using molecular and biotechnological techniques. The causative agents can be grouped according to the concentration of viruses shed by the patients as shown in Table 3. The seroconversion against the viruses is also given, since it indicates the efficacy of the hygienic preventive measures in our everyday life (i.e. the effectiveness of our everyday "Operation Clean Hands"). Through biotechnology, sensitivities of detection have been reached which had not been possible before; these data could not be obtained using classical technologies.

TABLE 3. Virus concentration in the fecal samples of patients, along with seroconversion data indicating the efficacy of preventive measures

Viruses	Concentration in stool samples [infecting particles per gram feces]	Location and age of sero-conversion of 50 % of inhabitants
Hepatitis A virus	10^5 - 10^8	Far East – school age
Polioviruses and other Enteroviruses	$10^5 - 10^8$ 10^5 - 10^8	Europe – 40-50 years of age
Rotaviruses	10^9 - 10^{12}	School age
Enteric adenoviruses	10^9 - 10^{12}	everywhere
Caliciviruses	10^9 - 10^{12}	USA, Europe,
Small-Round-Particles	10^9 - 10^{12}	Far-East

The Polymerase Chain Reaction (PCR) has proved to be one of the most useful tools for detection and identification of microorganisms that are not able to be cultured. The PCR is essentially a method for amplifying or multiplying DNA molecules in a test tube. In the matter of a few hours, specific regions of DNA molecules from an organism can be multiplied up to one billion fold. Theoretically, the method is able to identify a single microbiological agent in a sample containing a large number of other organisms [41], although usually 18-40 genome equivalents (18-40 cells or particles of that organism) are required for detection [42]. Combined with gel electrophoresis procedures that discriminate between PCR products according to characteristic fragment size and with the use of restriction endonucleases to produce restriction fragment length polymorphisms (RFLP) [31], PCR can be used to identify microorganisms. This is especially the case if nucleotide sequencing of the PCR product is also carried out. PCR can even be performed in some cases where the nucleotide sequence is not already known by random amplification of polymorphic DNA (RAPD) techniques [43]. Detection of toxin genes, resistance plasmids, and microorganisms used for decontamination of chemical agents [28] can be accomplished using these PCR- and RFLP-based methods.

Those who have used the PCR know that the chance of contamination of samples with PCR products (amplimers) is great because the concentration of the products is accordingly high. Contamination with amplimers is difficult to eliminate; therefore, a series of measures has been developed to prevent spillover of the products into other samples. The most recent achievements are the one-tube reverse-transcriptase (RT)-PCR and different isothermic techniques that are able to reduce the risk of contamination, since one can work without opening the vials after amplification. When genetic material is cloned into host cells, the probability of contamination is further reduced.

Further new techniques for the detection and identification of small, specific molecular differences in agents include high density oligonucleotide arrays and DNA chips [44]. Recent biochemical and biophysical techniques combined with laser beam fluorescence detection technologies have made it possible to identify very small arrays on a surface by computer technology. In this regard, minute regions of a surface are covalently covered by oligonucleotides of known nucleotide sequence, and these are used to identify unknown nucleotides through hybridization of the known sequences with the unknown samples. Originally, the technique was invented to detect mRNA sequences actively expressed by cells for the presence of very small amounts of Mycobacteria or HIV in diagnostic samples, but it can potentially be used for the detection of minor differences in the genomic DNA of biotechnologically manipulated organisms.

All of these techniques can be used in a positive way to investigate the possible use of BTW agents, for example, in potential manufacturing facilities, in patients hospitalized for unexpected illnesses [18], or animals with illnesses suspected to be caused by emerging infectious agents [6, 27].

4. Sources of Particular Concern in the Area of Virology

It has been mentioned above that in addition to biological warfare sorties and accidents with BW agents, bioterrorism might cause public health disasters in our era of globalization [14, 17, 21]. Accordingly, virologists have to call attention to some possibilities involving viruses which are cause for concern.

A prominent one is connected with the use of recombinant vaccinia virus-based oral vaccines, which have been used successfully for the prevention of rabies in wild animals (foxes, racoons) both in Europe and in the USA [33]. This demonstrates that vaccinia is a very successful vector for gene deliver, and therefore could be misused as a vector to deliver agents of BW relevance.

As discussed above, food-borne infections represent a global public health threat [20], and it has to be taken into account that such recombinants might be used as the means of terroristic attacks. Vaccinia virus recombinants expressing replication competent, positive strand RNA viruses might be sources of nation-wide disasters in a population of individuals in the 20s age group or younger, who are not vaccinated against smallpox (have not received immunization with the vaccinia virus). In this regard, vaccinia virus would not be a successful vector in a vaccinated population, as the immune system would probably eliminate the virus before it had a chance to deliver its recombinant gene package. Therefore, reagents able to detect traces of vaccinia recombinants should be included in diagnostic kits. Vaccinia virus vectors are available throughout the world, and dry aerosols as a means of delivery have been developed. The achievements of using vaccinia virus as a vector for therapeutic purposes have to be appreciated, but these are techniques that can definitely be misused for biological warfare or terroristic activities.

Fowl influenza viruses H5N1 (and probably H5N2) have been shown to infect human beings [38, 39]. Even classical virology might be able to produce dry aerosols to infect people [5]. Therefore, the BTW diagnostic package has to contain RT-PCR reagents to detect influenza viruses, specific for nucleoproteins as well as both subtype and variant-specific hemagglutinins and neuraminidases.

Hantaviruses appeared recently as a world-wide epidemiological problem [32, 9]. Arenaviruses are similar in the respect that no vectors are necessary for the human infections. These viruses are very stable in dry aerosols; therefore, one has to be prepared for the detection of such viruses with group-specific reagents [22, 23].

Transgenic plants might be also the vehicles of food-borne infections or intoxications. Several toxic substances have been introduced into transgenic plants up to now (scorpion and bacterial). Their detection is, however, identical with the techniques used for the detection of natural toxins and toxin genes.

Human and animal viruses which could be present in food-borne or aerosol-born infections are summarized in Table 4. This list has been compiled by the virologists working on emerging or reemerging viruses. Some of them are able to cause disease only for a limited proportion of the affected persons (enterovirus 71). It has been suggested, however, that the disease in one per cent of the population might result in the collapse of a public health system and health care. The preventive diagnostic package should contain specific reagents which are able to detect all of the above etiologic agents. The procedure of choice is the PCR, the ligase chain reaction (LCR) or one of the high density oligonucleotide array technologies. These techniques are the least dangerous for the laboratory personnel, since the procedures begin with the heat denaturation of the samples inactivating infectivity, too.

TABLE 4. Human viruses which might produce public health disasters if distributed in the form of recombinants or dry aerosols in connection with terroristic events [34].

Human viruses of enteral and/or respiratory transmission	Animal viruses of enteral and/or respiratory transmission
Enteroviruses: poliovirus (vaccine)/ EV70, 71 Human rotaviruses (vaccines) Human adenovirus serotypes (vaccines) Smallpox (vaccine) Influenza viruses	Hantaviruses (vaccine) Camelpox virus Monkeypox virus Horse morbillivirus Fowl influenza viruses H5N1 and H5N2 Crimean-Congo Ebola virus Argentinian hantaviruses Rabies (vaccine), Rodent-borne Arenaviruses, Herpesvirus simiae, Simian hemorrhagic fever, African swine fever, African horse sickness

(vaccine) indicates that a vaccine is available

In comparison to the ones listed in Table 4, some viruses are borne by insect vectors or carriers including Dengue (vaccine), Yellow Fever (vaccine), Chikungunya, Epidemic Polyarthritis, West Nile, St. Louis and Japanese B Encephalitis (vaccine), Kyasanur Forest, Rift Valley [10], Ross River [23], La Crosse, Eastern, Western and Venezuelan Equine Encephalitis (vaccine), Oropouche [13, 22, 23], Guanarito and Sabia viruses [20].

Dangers of biological terrorism might affect the health status of the population indirectly. The nature of different dangers is summarized in Table 5.

TABLE 5. Possible targets and sources of biological terrorism with agents considered to be the cause of emerging infections.

- Reinfection of vector populations, infection of new vectors [10], natural disasters [24].
- Spread of vector-borne viruses in ways other than through vectors, e.g. through blood products.
- Zoonoses transmitted in ways other than through animal sources.
- Vaccinia recombinants [33]. Other potential viral vectors to be identified as potential BTW agents.
- Fecal, waterborne and airborne transmission of agents, and interspecies transmission
- Sporadic infections and the potential collapse of health care and public health systems.
- Agents encapsulated in liposomes for uptake in skin cells, dendritic cells and other cells of the immune system [45].

The liposome technique is also an achievement of biotechnology. Industry utilizes it on wide scale in cosmetics. Biologically active substances can be introduced into skin cells, and subcutaneous immunologically active dendritic cells [45]. This potent technique unfortunately might also be a procedure which can be used for the introduction of infectious or toxic agents into the organism. This possibility has to be taken into account when diagnostic and preventive measures are planned.

5. Perspectives for the Future

The prospects for the detection and prevention of BTW agents will be improved in the future by the further development of DNA chip and high density oligonucleotide array procedures in addition to the PCR, LCR, fluorescence activated cell sorter (FACS) as well as other luminescence and photometric technologies [1, 4, 19, 29].

DNA chips will be developed to detect groups of viruses, and virus typing reagents will be developed which can detect the serotypes and genom types of possible infectious agents. DNA vaccines and combinations of mixtures of human monoclonal antibodies will be produced. New interferons, antiviral drugs, and new cytokines of recombinant origin will be available.

Nevertheless, one has to improve on possibilities of preventing production of BTW agents. In this regard, a global surveillance system will have to be introduced to detect both emerging viruses and other emerging BTW agents in due time for effective prevention.

6. References

1. Adam, P., Descroix, D., and Chiaroni, J-P. (1998) Flame photometry for biological detection, in *Proceedings of the 6th International Syposium on Protection Against Chemical and Biological Warfare Agents*, Stockholm, Sweden, 10-15 May, 61-68.
2. Bianchi, N.O. (1991) Chemical-biological warfare and aggression. Its origins and evolution, in Geissler, E. and Haynes, R.H. (eds.), *Prevention of a biological and toxin arms race and the responsibility of scientists*, Akademie Verlag, Berlin, 415-425.
3. Busbee, B.G. (1997) The technology of biological arms control and disarmament. Protection Technologies: Military and civilian. Paper presented at the NATO ARW "The technology of biological arms control and disarmament", Budapest 28-30 March.

4. Casadevall, A. (1996) Antibody-based therapies for emerging infectious diseases, *Emerging Infectious Diseases* **2**, 200-208.
5. Chanal, P.Y. (1998) The use of a low speed wind tunnel for evaluation of bio aerosol samplers. in *Proceedings of the 6th International Syposium on Protection Against Chemical and Biological Warfare Agents*, Stockholm, Sweden, 10-15 May, 83-88.
6. Childs, J., Shope, R.E., Fish, D., Meslin, F.X., Peters, C.J., Johnson, K., Debess, E., Dennis, D., and Jenkins, S. (1998) Emerging zoonoses, *Emerging Infectious Diseases* **4**, 453-454.
7. Collins, J.E. (1997) Impact on changing consumer lifestyles on the emergence/reemergence of foodborne pathogens, *Emerging Infectious Diseases* **3**, 471-479.
8. Conlon, D. (1998) Meeting summary on the Conference on global disease elimination and eradication as public health strategies, 26 February, 1998, *Emerging Infectious Diseases* **4**, 349-351.
9. Faludi, G. (1997) *Alteration of the importance of biological weapons* (in Hungarian), Budapest.
10. Fontenille, D., Traore-Lamizana,M., Diallo, M., Thonnon, J., Digoutte, J.P., and Zeller, H.G. (1998) New vectors of Rift Valley Fever in West Africa, *Emerging Infectious Diseases* **4**, 289-293.
11. Gao, S-J., Moore, P.S., Phil, M. (1996) Molecular approaches to the identification of unculturable infectious agents. *Emerging Infectious Diseases* **2**, 159-167.
12. Geissler, E. (1991) Biological weapons and the responsibility of scientists - twenty years later, in Geissler, E. and Haynes, R.H. (eds.), *Prevention of a biological and toxin arms race and the responsibility of scientists*, Akademie Verlag, Berlin, 3-28.
13. Gubler, D.J. (1998) Resurgent vector-borne diseases as a global health problem, *Emerging Infectious Diseases* **4**, 442-450.
14. Henderson, D.A. (1998) Bioterrorism as a public health threat. *Emerging Infectious Diseases* **4**, 488-494.
15. Jaykus, L-A. (1997) Epidemiology and detection as options for control of viral and parasitic foodborne disease, *Emerging Infectious Diseases* **3**, 529-539.
16. Jones, S.B., Unconventional pathogen countermeasures (Abstract). *Proc. 6th CBw Protection Symp.* (1998) Stockholm, 10-15 May, 1998, p. 161.
17. Kaufmann, A.F., Meltzer, M.I., and Schmidt, G.P. (1997) The economic impact of a bioterrorist attack: are prevention and post-attack intervention programs justifiable?, *Emerging Infectious Diseases* **3**, 83 - 94.
18. Knoke, J.D. and Gray, G.C. (1998) Hospitalizations for unexpected illnesses among U.S. veterans of the Persian Gulf War, *Emerging Infectious Diseases* **4**, 211-219.
19. Kulaga, H., Anderson, P.E., Cain, M.C., and Stopa, P.J. (1998) Detection and identification of biological agents by flow cytometric analysis, *Proc.*

6th CBw Protection Symp., Stockholm, 10-15 May, 1998, 95-99.
20. Lederberg, J. (1997) Infectious disease as an evolutionary paradigm. *Emerging Infectious Diseases* **3**, 417-423.
21. McDade, J.E and Franz, D. (1998) Bioterrorism as a public health threat, *Emerging Infectious Diseases* **4**, 493-494.
22. Murphy, F.A. and Nathanson, N. (1994) The emergence of new virus diseases: an overview. *Seminars in Virology* **5**, 87-102.
23. Murphy, F.A. (1998) Emerging zoonoses, *Emerging Infectious Diseases* **4**, 429-435.
24. Nasci, R.S. and Moore, C.G. (1998) Vector-borne disease surveillance and natural disasters, *Emerging Infectious Diseases* **4**, 333-334.
25. Nixdorff, K. (2000) The biotechnology revolution: the science and applications, in G.S. Pearson and M.R. Dando (eds.), *New Scientific and Technologic Aspects of Verification of the Biological and Toxin Weapons Convention (BTWC)"*, Kluwer Academic Publishers, Dordrecht, forthcoming.
26. NSTC-CISET Working Group on Emerging and Reemerging Infectious Diseases Infectious disease - a global health threat. Washington (DC): U.S. Government Printing Office, September 1995
27. Philbey, A.W., Kirkland, P.D., Ross, A.D., Davis, R.J., Gleeson, A.B., Love, R.J., Daniels, P.W., Gould, A.R., and Hyett, A.D. (1998) An apparently new virus (family *Paramyxoviridae*) infectious for pigs, humans and fruit bats, *Emerging Infectious Diseases* **4**, 269-271.
28. Rajan, K.S. and Mainer, S.: Exploration of microbial sources for CW-agents destruction. Proceedings of the 6th CBW Protection Symposium, 349-355, Stockholm, 10-15 May, 1998.
29. Rolland, X., Herbig, E., Guénec, O., Icke, B., Brailsford, M., and Drocourt, J.L.: ChemScan-RDI: A real time and ultra-sensitive laser scanning cytometer for microbiology. Applications to water, air, surface and personnel monitoring. Proc. 6th CBw Protection Symp. pp103-110, Stockholm, 10-15 May, 1998.
30. Russo, M., Burgyán, J., and Martelli, G.P. (1994) Molecular biology of *Tombusviridae, Adv. Virus Res.* **44**, 381-428.
31. Salamon, H., Segal, M.R., de Leon, A.P., and Small, P.M. (1998) Accommodating error analysis in comparison and clustering of molecular fingerprints, *Emerging Infectious Diseases* **4**, 159-168.
32. Schmaljohn, C. and Hjelle, B. (1997) Hantaviruses: a global disease problem, *Emerging Infectious Diseases* **3**, 95-104.
33. Steck, F., Wandler, A., Bischel, P., Capt, S., Hafliger, U., and Schneider, L. (1982) Oral immunization of foxes against rabies. Laboratory and field studies, *Comp. Immunol. Microbiol. Infect. Dis.* **5**, 165-179.
34. Stephenson, J. (1996) Confronting a biological Armageddon: experts tackle prospect of bioterrorism, *JAMA* **276**, 349-351.
35. Thränert, O. (1991) The United States unilateral renunciation of biological and toxin warfare agents, in Geissler, E. and Haynes, R.H. (eds.), *Preven-*

tion of a biological and toxin arms race and the responsibility of scientists, Akademie Verlag, Berlin, pp. 129-140.

36. Tucker, J. (1997) Biological weapons proliferations concerns, in G.S. Pearson and M.R. Dando (eds.), *New Scientific and Technologic Aspects of Verification of the Biological and Toxin Weapons Convention (BTWC)*", Kluwer Academic Publishers, Dordrecht, forthcoming.

37. Valdes, J.J. (1997) Biological agent detection technology, Paper presented at the NATO ARW "The technology of biological arms control and disarmament", Budapest 28-30 March.

38. Webster, R.G. and Kawaoka, Y. (1994) Influenza - an emerging and re-emerging disease, *Seminars in Virology* 5, 103-112.

39. Webster, R.G. (1998) Influenza: an emerging disease, *Emerging Infectious Diseases* 4, 436-441.

40. Zilinskas, R.A. (1991) Biological warfare and the third world, in Geissler, E. and Haynes, R.H. (eds.), *Prevention of a biological and toxin arms race and the responsibility of scientists*, Akademie Verlag, Berlin, pp. 183-211.

41. Amann, R.I., Ludwig, W., Schleifer, K.-H. (1995) Phylogenetic identification and in situ detection of individual microbial cells without cultivation, *Microbiological Reviews*, 59, 143-169.

42. Cohen, H.J., Mechanda, S.M., and Lin, W. (1996) PCR amplification of the finA gene sequence of *Salmonella typhimurium*, a specific method for detection of *Salmonella* ssp., *Applied and Environmental Microbiology*, 62, 4303-4308.

43. Welsh, J., McClelland, M. (1990) Fingerprinting genomes using PCR with arbitrary primers, *Nucleic Acids Research*, 18, 7213-7218.

44. Lockhart, D.J., Dong, H., Byrne, M.C., Follettie, M.T., Gallo, M.V., Chee, M.S., Mittmann, M., Wang, C., Kobayashi, M., Horton, H., Brown, E.L. (1996) Expression monitoring by hybridization to high-density oligonucleotide arrays, *Nature Biotechnology* 14, 1675-1680.

45. Ludewiga, B., Barchiesia, F., Pericina, M., Zinkernagela, R.M., Hengartnera, H., Schwendenerb, R.A. (2000) In vivo antigen loading and activation of dendritic cells via a liposomal peptide vaccine mediates protective antiviral and anti-tumour immunity, *Vaccine*, 19, 23-32.

MEDICAL COUNTERMEASURES TO BIOLOGICAL WARFARE AGENTS

DAVID R. FRANZ
Southern Research Institute
431 Aviation Way
Frederick, MD, 21701 USA

1. Biological Warfare

For nearly 50 years, NATO nations have developed biological weapons or countermeasures to biological weapons [1], both efforts pursued in response to the threat of possible first use of biological warfare agents against the alliance. Prior to the termination of the biological warfare programs of NATO members in the late 1960s, an offensive capability was considered the best defense or deterrent against Soviet use of biological warfare. The focus of biological defense programs, then and now, has been protection of military forces. The major world powers developed field detectors, masks and protective clothing, vaccines, drugs and diagnostics to eliminate or reduce the potential threat. Although it is now known that the US, Russia, the UK, Canada and other nations possessed biological weapons, the threat has not been realized on a modern battlefield.

In the last ten years of the 20th century, technological and political factors changed the face of biological warfare forever. In 1989, the world saw a decisive engagement of the Iraqi military by an overwhelming, technologically superior coalition force. The US and its allies demonstrated that there was no match for laser-guided bombs and GPS guided cruise missiles on the conventional battlefield. And, for the first time, the world watched the war in real time on television. Observing the conventional might of the coalition, one can assume that potential adversaries began thinking about the need for a "great equalizer" to counter the demonstrated superiority.

Immediately after the war in the Gulf, the Soviet Union collapsed; we were to learn later that the Soviets were without peer in the field of offensive biological warfare. The enormous infrastructure built by the USSR for the research, development, production and testing of tactical and strategic biological weapons was now split between the newly independent states and began to crumble. Defense funding of many facilities designed and built for the production of bio-

A. Kelle et al. (eds.), The Role of Biotechnology in Countering BTW Agents, 227–234.
© 2001 Kluwer Academic Publishers. Printed in the Netherlands.

logical weapons slowed to a trickle. Concern was noted in the West that highly skilled BW scientists and engineers, now unemployed, might be tempted to sell their expertise to other nations seeking to level the playing field with superior conventional capabilities.

2. The Biological Revolution

While the Soviet Union stumbled and lesser nations combined radical ideology with a commitment to counterbalance the West's conventional advantage, the revolution in biotechnology accelerated. The human genome project, genetic engineering, and recombinant technologies now promise to make our lives better, but at the same time remind us that, historically, beneficial technological developments are sometimes exploited for evil as well. Non-biological terrorist incidents close to home have raised our awareness and concern that a new generation of biological terrorists might use bacteria, viruses, toxins or even designer agents against our forces or our cities [2].

Throughout the cold war, proliferators had independently selected nearly the same list of weapons agents for tactical use on the battlefield or strategic use to be delivered by intercontinental ballistic missiles. It was a relatively small group (ca. 10-20) of agents drawn from hundreds available in nature that had the physical and biological characteristics such as pathogenicity, toxicity, ease of production and stability qualifying them for aerosol dissemination. Thus, the cold-war "threat" list(s) was developed [3]. Today, as bioterrorism – biological attacks targeting civilians not directly associated with armed conflict – has become a concern, generally the same constraints face the terrorist selecting agents for the production of biological weapons. To produce a true mass-casualty attack the terrorist will likely have to use agents from that threat list developed and tested for biological warfare application. However, because a small non-lethal attack – or even a biological hoax – is adequate to cause panic and lead to coveted national television coverage, the modern bioterrorist probably has a much wider spectrum of agents from which to choose. However, the terrorist spectrum, while much broader, is not necessarily more lethal. The large community outbreak of salmonellosis caused by intentional contamination of salad bars in the U.S. state of Oregon in 1984 serves as an example of an effective non-lethal attack [4].

Today's challenge is to utilize the enormous advances in biotechnology for the protection of our military forces and civil populations from the very threats that such progress has served to facilitate.

3. Vaccines

Medical countermeasures to biological weapons have traditionally included vaccines, chemotherapeutic drugs and diagnostics [5]. Vaccines played the central role in our research and our force protection programs throughout the cold war. They provide long-term protection, are effective against most of the threat agents of concern, and are generally difficult to circumvent. Because they can be given prophylactically and provide effective protection for years, vaccines are logistically the most efficient means of protection of military forces, at least until real-time stand-off detection systems become a reality. Physiologic limitations of the human immune system make it necessary to immunize individuals weeks before potential exposure; this fact necessitates articulation and implementation of an immunization policy well before the onset of hostilities. Additionally, most biodefense vaccines have limited utility in the civilian public health arena; therefore, market factors have not stimulated interest by industry. Finally, biodefense vaccines are unique in that, because there are not naturally-occurring outbreaks of most of these diseases, human efficacy testing can typically not be done, making licensing difficult in most countries.

Advances in biotechnology are facilitating development of vaccine candidates for biodefense in the laboratory and eliminating constraints such as the requirement for biocontained production laboratories. It is likely that these new technical options and methods will also reduce regulatory compliance difficulties in the future. An example is the recent development of alphavirus replicon vaccine "carrier" technology. A version of this vaccine system is based on an avirulent recombinant Venezuelan Equine Encephalomyelitis (VEE) virus into which DNA coding for bacterial, toxin or other viral antigens can be inserted for expression by the carrier virus after injection into the recipient. Because VEE is a lymphotrophic virus, the vaccine system targets the immune system directly, thus extremely efficiently. Because this unique, nonreplicating carrier could be used to deliver a large number of vaccines by changing only the nucleic acid portion coding for the specific antigen, regulatory approval may be easier in the future.

4. Antibiotics

Bacterial agents contain their own genetic material which allows them to multiply outside living cells, unlike viruses, which are dependant on host (animal or plant) cells to provide genetic machinery for replication. Bacterial metabolic schemes allow intervention by blocking cell wall synthesis, protein synthesis or even uptake of nutrients by the organism. Antibiotics exploiting these targets are effective against many bacterial agents, if given before clinical signs of disease. Antibiotics are readily available in most countries because of public health market demand. Unfortunately, they provide only single incident protection and require identification and/or sensitivity testing of the agent against which they are to be used. Furthermore, it is relatively easy to engineer bacterial strains – even in a low-tech environment – to be resistant to single or multiple antibiotics. Finally, it is little appreciated that the gastrointestinal and other systemic side effects of taking certain antibiotics prophylactically for several weeks may cause significantly more discomfort than the short-lived effects of most immunizations. Antibiotics are not a panacea; they are used most effectively as an adjunct to vaccines as part of an integrated package to deal with unknown or unexpected bacterial agent exposures where there is not time to immunize or no vaccine exists. Biotechnology – and now the urgency provided by natural selection of resistant bacterial strains important to public health – will facilitate the development of new antibiotics. Molecular biology can help us understand, and overcome, the mechanisms of microbial resistance and market factors will drive research and development. Now, microencapsulation – a technology that allows both timed release and direct targeting of specific cell types – of already available antibiotics may make them more potent in treating certain intracellular threat microbes. It is unlikely that specific programs will be necessary for "biodefense antibiotics" because of the broad civilian application of this class of countermeasures.

5. Antiviral Drugs

Unlike most bacteria, viruses are intracellular pathogens. In order for a single viral particle to duplicate itself, it must bind to the outer membrane and enter its host cell, become uncoated, replicate its nucleic acids and produce viral proteins using host machinery, then reassemble and bud from the primary host cell to infect another. Each of these processes provides a target for antiviral drugs, typically via the enzymes involved. Probably because viral replication is so intimately entwined with normal cell function – and because viruses are intracellular parasites – developing drugs that kill them without damaging their host cells has been difficult. Thus, there are very few antiviral drugs available. Like antibiotics, antivirals must be given before signs of disease for best result. Because most viral – and bacterial – syndromes begin with flu-like symptoms, timely definitive diagnosis is difficult. While their intracellular life-styles and dependent nature make the viruses difficult chemotherapeutic targets, those same characteristics probably make the natural or malicious development of drug resistance more difficult as well.

Defense research programs have recently begun leveraging the well-funded research into anti-HIV drugs and thereby some progress is being made to broaden our chemical armamentarium against viral threat agents. Viral genomics will obviously increase our understanding of the target, thus focus the development of new antivirals. Dissecting the genomes of viral threats and learning their vulnerabilities may also lead us to entirely new classes of drugs which target more general pathogenic mechanisms such as the complex cytokine cascades, critical to many other disease manifestations. A good example of this emerging possibility is the recent discovery of proteins, produced by the smallpox virus, which appear to effect specific cytokine-receptor interactions.

6. Antibody Preparations

Antisera have been used for years to treat certain diseases or intoxications. The botulinum toxins are probably the most familiar BW threat agents for which antisera are available [6]. Polyclonal antisera have the major disadvantage of requiring large-scale production in a species other than man, typically the horse. Therefore, unless the antibody preparation is despeciated, humans treated with these heterologous antibodies are likely to suffer an allergic reaction to the foreign protein. Small quantities of human antisera have been produced for the botulinum toxins and for vaccinia (smallpox vaccine) in the past. This was done by repeatedly immunizing a population of humans, then harvesting their antibodies by a process called plasmapharesis. Such antibodies, when given to another human, may remain in circulation for weeks, rather than days for animal antibodies in humans, because they are not recognized as foreign. The high incidence of HIV, hepatitis and potentially yet unknown viruses in the donor population has made mass production of human-derived antisera nearly impossible in some countries from a regulatory and safety standpoint. Biotechnology has now made the production of human monoclonal antibodies feasible [7]. Human monoclonal antibodies are natural or synthetic antibodies which have an affinity for a single antibody recognition site on a bacteria, virus or toxin, whereas polyclonal antibodies, like those produced by a response to a vaccine, are essentially multiple (poly) clones of (monoclonal) antibodies. Like vaccines and drugs, antibody preparations are not a total solution, but may be of value as part of an integrated approach to medical countermeasures. Monoclonals are presently very expensive and typically not as effective as polyclonals. However, they do offer promise for certain viral diseases, potentially providing passive immune therapy or immediate prophylactic protection without the 4-6 week wait associated with active immunization with vaccines. Recent breakthroughs in monoclonal antibody discovery and production have been facilitated, for example, by the development of technologies which allow us to harvest lymphocytes from an immunized human, fuse those lymphocytes with specially designed cells in vitro, and produce human antibodies which neutralize antigen and should not be recognized as foreign when administered to humans.

7. Rapid Diagnostics

Today, a military force might be immunized and sent into harm's way with protective masks and blister packages of antibiotics in their packs. The accompanying medical laboratory unit will have modern diagnostic tools in its kit. Detectors might warn of an attack and maybe even identify the agent within an hour of its release. The biological terrorist threat to our less well-prepared cities – or our military forces in garrison – may be even more difficult to manage than the classical battlefield threat. In fact, bioterrorism may be very difficult to differentiate from a naturally occurring infectious disease. Thus, the first response opportunity may be identification of the agent *after* the attack on a inexpectant population. In this scenario, a rapid means of definitive identification and triage of patients exposed, but not yet showing signs of illness, may be the most important means of reducing loss of life [8]. Antibody based diagnostics have now been fielded for many of the agents of concern. Nucleic acid based assays (e g. polymerase chain reaction), with greatly increased sensitivity, are available in reference laboratories throughout the world and miniaturized platforms have been tested in field settings. Recombinant technologies have also made reagent preparation easier and more efficient. Therefore, because regulatory constraints are less rigid for diagnostics than for vaccines or drugs, the powerful impact of technology is already being felt in the field.

8. Medical Countermeasures of the Future

The phenomenal advances in biotechnology will only enhance our ability to diagnose and medically protect or treat our military and civil populations following a biological warfare attack, whether on the battlefield or on the street corner. Agents will be identified using lightweight hand-held devices that amplify and identify gene targets in the field. Mass spectroscopy and microchip technologies will provide rapid, much less reagent-intensive systems for field identification. Cheaper, cleaner, more rapidly developed vaccines will result as infectious clone, replicon, recombinant peptide and DNA vaccine technologies mature. Computer aided structural analysis and high-throughput, robotic, mechanism-based assay systems will lead to more affordable and effective antibiotics and antiviral drugs [9].

Protecting our citizens from the future biological threat will require an integrated approach and teamwork among the law enforcement, public health, emergency response, military and intelligence communities. Application of the latest tools for identification and protection through exploitation of the rapidly advancing biotechnologies will facilitate the development of better medical countermeasures [10]. What we know about the threat of biological agents and weapons can truly make a difference in the way we react and respond. Therefore, we must educate our health-care providers, our first responders, our citizens and our local and national leadership. Emerging electronic communication technologies are already impacting this vital part of preparation. Understanding the threat, good intelligence, a well conceived passive defense, robust military systems to directly target an aggressor bent on using biological agents, and treaties designed to strengthen the international norm against the development and use of biological weapons can serve not only as defense, but as a deterrent by raising the risk and reducing the benefit to the would be bioterrorist or weaponeer. Economic analysis of a bioterrorist attack on a civilian population suggests that the long-term costs could be as much as $477.7 million for a brucellosis attack and $26.2 billion for an anthrax attack per 100,000 persons exposed [8]. On January 22, 1999, President Clinton proposed an additional, and unprecedented, $1.4 billion for research, drug and vaccine stockpiles, education, training and surveillance programs to defend against domestic bioterrorist attack. These initiatives serve an important dual-use by buttressing our public health infrastructure. This proposed investment in preparation and deterrence, costing a fraction of the potential cost of a successful attack against our cities, must leverage the enormous biotechnological breakthroughs of the last 20 years. Tomorrow's breakthroughs in biotechnology, which might also be used to improve old weapons or develop new ones, if fully exploited, will likely be much more useful and effective in protecting freedom-loving nations from those who would use biological weapons against us.

9. References

1. Christopher, G.W., Cieslak, T.J., Pavlin, J.A., and Eitzen, E.M. (1997) Biological Warfare: A Historical Perspective, *J. American Medical Association* **278**, 412-417.
2. Carus, W.S. (1998) *Bioterrorism and Biocrimes: The Illicit use of Biological Agents in the 20th Century*, Working Paper; Center for Counterproliferation Research, National Defense University, Washington, D.C.
3. World Health Organization (1970) *Health Aspects of Chemical and Biological Weapons: Report of a WHO Group of Consultants*, Geneva, Switzerland, pp. 98-99.
4. Torok, T.J., Tauxe, R.V., Wise, R.P., Linvengood, J.R., Sokolow, R., Mauvais, S., Birkness, K.A., Skeels, M.R., Horan, J.M., and Foster, L.R. (1997) A Large Community Outbreak of Salmonellosis caused by intentional contamination of restaurant salad bars, *J. American Medical Association* **278**, 389-395.
5. Franz, D.R., Jahrling, P.B., Friedlander, A.M., McClain, D.J., Hoover, D.L., Bryne, W.R., Pavlin, J.A., Christopher, G.W., and Eitzen, E.M. (1997) Clinical Recognition and Management of Patients Exposed to Bio-

234

logical Warfare Agents, *J. American Medical Association* **278**, 399-411.

6. Franz, D.R., Pitt, L.M., Clayton, M.A., Hanes, M.A., and Rose, K.J. (1993) Efficacy of prophylactic and therapeutic administration of antitoxin for inhalation botulism, in B.R. DasGupta (ed.), *Botulinum and Tetanus Neurotoxins*, Plenum Press, New York, pp. 473-476.

7. Brownlee, S. (1995) Biotech finally finds the bottom line, *U.S. News and World Report*, July 17, p. 45.

8. Kaufmann, A.F., Meltzer, M.I., and Schmid, G.P. (1997) The economic impact of a bioterrorist attack: Are prevention and postattack intervention programs justifiable, *Emerging Infectious Diseases* **3**, 83-94.

9. Major, J. (1998) Challenges and opportunities in high throughput screening: Implications for new technologies, *J. Biomolecular Screening* **3**, 13-17.

10. Rosen, P. (1999) *Chemical and Biological Terrorism: Research and Development to Improve Civilian Medical Response,* Institute of Medicine, National Research Council, National Academy Press, Washington, D.C.

TREATMENT OF MASS CASUALTIES UNDER WORST CASE ASSUMPTIONS

SVEN-ÅKE PERSSON
Swedish Defence Research Establishment (FOA)
Department of NBC Defence
S-901 82 Umeå, Sweden

1. Introduction

The great plague pandemic in Europe that lasted from 1346 to 1352 is an example of a classic worst case scenario. The well-known clinical presentations were bubonic, primary septicaemic and pneumonic disease [1]. It spread through Europe and killed 25 million people or about one quarter of the European population. We know of other devastating pandemics of bubonic plague throughout history [2]. This "Black Death" was one of at least three known pandemics. Most recently, a suspected outbreak of bubonic plague occurred in the Indian town of Surat in 1994. Fearing the disease, especially the often fatal pneumonic form, citizens including medical personnel tried to escape the afflicted area. However, despite a significant number of infections there were only a small number of fatal cases. There are still areas in Africa, America and Asia where bubonic plague is endemic, causing a mortality rate of approximately 10-20 % in contrast to some 60 % during the medieval pandemics.

In 1918-1919 Sweden, among other countries, was plagued by the pandemic "Spanish Flu", which not only resulted in significant morbidity but also significant mortality [3]. It has been estimated that 20 to 50 million people lost their lives in this pandemic. No specific treatment was at hand. Birch sap mixed with brandy was the only remedy available.

Which scenarios are today the worst case assumptions? What measures could be taken? How should we treat mass casualties?

2. Political Factors That Could Affect a Scenario

We are living in a world with yawning chasms between developed countries with resources and poor countries demanding development. Water and food

A. Kelle et al. (eds.), The Role of Biotechnology in Countering BTW Agents, 235–245.
© 2001 *Kluwer Academic Publishers. Printed in the Netherlands.*

resources and good health care are also unevenly distributed in favor of rich and developed countries. Ethnic conflicts in the Balkans, unresolved conflicts between Israel and Palestine and difficulties surrounding Iraq's cooperation with the United Nations Special Commission are a few examples of potential trouble spots. Recently there has been serious instability in the global economy, with major difficulties in Asian countries such as Indonesia, Cambodia and Malaysia causing unemployment, poverty and political instability. Nor should we forget the instability and economic problems of Russia. Relations between India and Pakistan are particularly strained after their testing of nuclear devices. North Korea has demonstrated missile capability to the world and Japan in particular, simultaneously increasing its isolation by refusing even medical assistance for its people. The Islamic fundamentalist regimes in Iran and Afghanistan support and encourage Muslims in other countries. Even if these countries are not directly involved in terrorist attacks or other activities of militant groups and terrorists, close connections with such groups have been disclosed on several occasions.

In recent years groups of environmental terrorists and "militant vegans" or animal rights activists, e.g. the Animal Liberation Front (ALF), have been increasingly active. Unfortunately their actions are often undemocratic, violent and sometimes illegal.

Chemical and Biological Warfare agents were used in Japan in 1994-95 by the sect Aum Shinrikyo [4]. The dreadful attack with the nerve agent sarin in the Tokyo subway is a warning. This frightening example could attract future followers.

Major recent advances in biotechnology and pharmaceutical sciences offer a large number of possibilities for new drugs and new principles for treating diseases. However, they also open the door for misuse of these techniques in creating engineered biological agents for warfare, sabotage or other harm. Production is cheap and comparatively simple and some formulae can easily be obtained through the Internet.

3. Possible Scenarios

A worst case scenario in which use of BTW agents produces massive casualties depends on where this use takes place. Use of BTW agents in a developed society with good infrastructure and well functioning systems for medical prevention and health care will probably cause serious problems, but much worse consequences could follow from BTW use in a less developed country with poor infrastructure and primitive systems for medical prevention and health care. Climate is another important factor for the outcome of the possible use of BTW. Therefore it is advisable to present worst case scenarios that differ from each other in these respects. I have selected Sweden, my own country, as "location" for the first two worst case scenarios. Sweden is a developed country with good infrastructure and well-developed systems for medical prevention and health care. In the third scenario I have tried to describe the effects of BTW use under much less favorable conditions.

4. Use of BTW Agents in a Developed Country – Two Worst Case Scenarios

4.1 BACKGROUND

Sweden has about 9 million inhabitants, widely but not uniformly distributed over a large area. Parts of the northern mountain regions are very sparsely populated. The majority lives in the densely populated area Stockholm region, near the city of Gothenburg and in the south of Sweden, close to Denmark. The level of education is high. Sweden has been subject to the negative economic developments of recent years and has a comparatively high level of unemployment. Sweden has a large export industry (cars, trucks, the timber industry, mobile telephones). We are food producers but we also import food. Food is stored to a limited extent and extensive transportation is needed to supply all parts of the country. Sweden has very large resources of drinking water. The energy and heat necessary for industry and for heating houses are produced by hydroelectric and nuclear power plants. The Swedish government strongly supports measures to improve the environment and make Sweden an ecologically sustainable society. Our comparatively large defense force has been reduced, but will be more professional and will also be given some new tasks, e.g. participation in international peace-keeping or peace-making operations or other activities such as the Partnership For Peace (PFP). However, Sweden is not a member of NATO.

In an international perspective Sweden has a well-developed civil defense. The National Rescue Services Board has been engaged in a number of relief operations abroad.

In case of big accidents or disasters local fire and ambulance services, police and hospitals can work together and are often trained to do so. Every hospital has a disaster plan that often also includes measures to be taken in case of a large chemical accident or in case of exposure to chemical warfare agents, for example after a terrorist attack. To increase the preparedness to handle chemical casualties without serious contamination of the ambulances or hospitals, the National Board of Health and Welfare has supported the emergency hospitals economically so that they can establish permanent or mobile decontamination facilities within or near the hospitals. Resources have also been allocated to develop comfortable, personal, protective equipment and to train the personnel in effective and safe decontamination procedures. Less attention has been given to problems that could appear in case of possible use of BTW agents. Thus we usually have to resort to the general measures taken to prevent infectious diseases. However, in Linkoping and Stockholm a special force, equipped and trained to take care of casualties infected by unknown and virulent agents, has been established.

All infants in Sweden are vaccinated against tetanus, diphtheria and now, after a 20 years halt, even against whooping-cough and also against measles, German measles, chicken parotitis and poliomyelitis.

The county medical officer responsible for infectious diseases in every county of Sweden is responsible for monitoring developments in the panorama of infectious diseases in his district and taking the necessary legal measures to prevent further transmission of infection, for instance by isolating carriers.

All outbreaks of infectious diseases in Sweden are reported to the Swedish Institute for Infectious Diseases Control.

4.2 BTW AGENT USE SCENARIOS

4.2.1 The background

Almost every year there are outbreaks of gastrointestinal infections in Sweden. They usually pass without very much notice until the number of cases is large, since only about 10% of those infected visit a doctor. When diagnosis of these usually waterborne infections has been established the pathogens have been *E. coli,* toxigenic *Campylobacter* and *Salmonella* or other agents.

Regular epidemics caused by the influenza virus also occur. The severity of these epidemics can vary from one year to another. In 1998 there were an unusually large number of small rodents, which is sometimes associated with a higher incidence of haemorrhagic fever with renal syndrome. In recent years a few cases of fulminating infections caused by extremely virulent streptococci have occurred. The use of antibiotics in pig and cattle feed can increase the risk of a number of infections. A single case of fatal infection after consumption of meat probably caused by antibiotic resistant salmonella bacteria has recently been reported in Denmark.

4.2.2 The situation in crisis and wartime

In crisis and wartime situations there will be a gradual decline in services provided to society due to increasing demands with decreasing resources. The production of energy and heat will be reduced. Room temperature will have to be lowered as will the temperature of hot water, a very severe situation in a country where the outdoor temperature can be several degrees below zero for months. Problems with supplies of oil and gasoline could seriously disrupt the transport of important goods. From time to time there could be difficulties in providing enough and drinkable water. The supply of food and drugs, which today are not very sustainable, could be seriously disrupted leading to rationing and to shortages of important medicines such as antibiotics. Even if a country is not directly involved in a war, there is a serious risk that the country will have to receive and accommodate a great number of refugees under poor hygienic conditions. The refugees could also be infected and/or disease carriers.

It is also known from detailed Finnish and Swedish material on infectious diseases recorded over a period of almost 30 years that an increase in the incidence of infections of the respiratory tract, gastrointestinal infections and venereal infections is to be expected under such crisis conditions. This will probably also be observed in future crisis or war situations. However, the living conditions of today differ in many aspects from those fifty years ago. Life today is much more comfortable than fifty years ago, but at the same time may also be more vulnerable. How will we manage to live under the primitive, uncomfortable and stressful conditions of war? How will our immune defense combat infections caused by a blend of new and old pathogens in situations where the availability of antibiotics could be much less than that required. The effects of psychological stress on resistance to infections are certainly important, but not very well known.

Because of the use or misuse of antibiotics, bacterial strains resistant to antibiotics have developed. There are risks that such pathogens could be transmitted by refugees or by animal food products, especially when there are insufficient resources for food control. Serious problems could appear in treating these infections because the non-resistant exclusive antibiotics needed could be very difficult to obtain in case of crisis and war. Furthermore it can be expected that infections caused by a new type of virus against which there is no known vaccine and no immune defense exists in the exposed population could, like the "Spanish Flu", produce high morbidity and high mortality.

4.2.3 Mass casualties under worst case assumptions

A war situation in Europe in which BTW is used is not very likely but cannot be excluded. The use of missiles equipped with BTW warheads appears fairly unlikely. Very few actors would be able to launch such a weapon effectively. The probability of Sweden being selected as the target for a missile attack is also very low. Sabotage or a terror attack seems to be a more credible scenario. The first one I have chosen is a scenario where a group of terrorists have decided to use virulent pathogens as a means to reach their political aims. The pathogen was either *Bacillus anthracis* or *Yersinia pestis*. Table 1 shows some important characteristics of the two microorganisms. The two pathogens have incubation times ranging from one day to a few days (Table 1, see end of chapter).

The terrorist group also decided to deliver the agent in a form to be inhaled. An aerosol ought to be a good choice. However, in selecting the agent to be used they will face some problems (Table 2, see end of chapter).

4.2.4 Scenario 1: Terrorist Attack

Three members of a terrorist group suspected of acts of terror in countries other than Sweden are arrested by national security police. Very soon the terrorist group announces that unless those arrested are released and granted safe conduct out of the country there will be acts of terror in Sweden. A message saying that extremely contagious diseases will be transmitted in a Swedish city is delivered to a television company. The police and security police investigate and discuss measures to be taken with other authorities (National Board of Health and Welfare (SOS), the county medical officer responsible for infectious diseases, Swedish Institute for Infectious Diseases Control (SMI), Defence Research Establishment (FOA)). These contacts are informal to avoid panic and flight, and the discussions are on topics such as protection, preparedness, information etc.

About one week later the police receive a message that a sample containing dangerous contagious matter has been placed in a safety-deposit box in the railway station. The police contact the SOS, SMI, the Emergency Force of SOS (EFSOS), FOA and the National Veterinary Institute (SVA).The station is cordoned off, and the fully protected police together with the EFSOS remove the box and the sample. The sample is delivered to FOA and SMI for analysis, which will take at least 24 hours. On the next day the terrorists inform the media that the sample contained bacteria causing plague. They also send a message to the police saying that plague bacteria will be released in the inner city unless their demands are met. FOA and SMI confirm that the sample contains *Y. pestis* and that its sensitivity to antibiotics has been examined.

Negotiations with the terrorists are now considered unavoidable. A number of other problems have to be dealt with:

- How to give clear and correct information to the population?
- How to reassure people in order to avoid panic and flight?
- Where are the stores of antibiotics located? Are they sufficient? If not, where to get further antibiotics? How to administer the distribution and administration of the antibiotics?

On the next day the television station gets a message that a bag continuously spraying bacteria has been placed within the central station. The police have to locate it, evacuate the station and cordon off the area. All incoming trains must be stopped. The passengers have to be informed and given antibiotics as a prophylaxis. The emergency plans against epidemics have to be checked and put into operation. A number of other serious and immediate problems must be solved:

- Where to care for those infected?
- How to protect the personnel?
- How to find and isolate disease carriers?
- How to deal with contagious material?
- How take care of the dead and how to bury them?
- How to take care of the victims' relatives?

A more long-term challenge relates to the question of how to deal with the aftermath of the psychological trauma those involved will suffer for years.

It can be estimated in the present scenario that about 5,000-10,000 persons could be infected if the plague-spraying device was allowed to spray for 30 minutes. But these numbers and the number of fatal cases will depend very much on the effectiveness of the countermeasures taken.

4.2.5 Scenario 2: Newly Emerging Disease

This is a scenario in which a new "Spanish Flu" emerges. During the late autumn of 1997 a "bird flu" emerged in Hong Kong, infecting 18 people and killing six [6]. Since December 28, 1997 no new cases have been discovered. The pathogen was a new, highly contagious and lethal virus. This "super virus" was created by an influenza virus originating in chickens and transmitted to humans. In man this chicken flu virus can combine with ordinary human strains. If someone is infected with both viruses at once, the RNA from one virus could reconfigure with the RNA from the other and create a new virus that is both deadly and contagious.

The avian H5N1 virus which caused the influenza has been characterized [7, 8]. It is very interesting that in a dead victim of the "Spanish Flu" with virus pneumonia, genes have also been isolated from the deadly virus suspected of causing the disease. The findings were consistent with a novel H1N1 influenza A virus belonging to a subgroup of strains that infect humans and swine [9].

A comparison between some characteristics of the "Spanish Flu" and a future influenza pandemic is shown in Table 3 (see end of chapter).

A virus, if not killed by the immune system of the host, can remain latent for a long time after infection and under favorable conditions can become active again. As discussed above, the properties of the virus could change via a process of recombination. However, recent advances in molecular biology have made tools available which make it possible to construct new viruses with properties different from those already known [10, 11]. These findings make it possible to develop new, target specific treatment, for instance to destroy cancer cells. Unfortunately this technique can also be used to deliberately create viruses to cause harmful effects. This Hong Kong incident shows the importance of an infectious diseases surveillance net. In this particular incident cooperation between U.S. Centers for Disease Control and local experts and experts from other countries was successful. If this had not been the case the outcome might well have been a pandemic with very high morbidity and mortality. This would have placed a heavy burden indeed on every country affected, including those with good resources for disease prevention and medical care.

5. Use of BTW Agents in a Developing Country – Worst Case Assumptions

What would be the result if BTW agents were used in countries such as Bangladesh, parts of India and countries in Africa e.g. Rwanda, Congo, Angola? Unfortunately, such use can be expected to make a bad situation even worse.

Under the circumstances pertaining in these countries it would be very difficult to obtain evidence of the use of BTW agents. For instance, the widespread incidence of enteritis caused by *Vibrio. cholerae* in Bangladesh and India would disguise any man-made infection caused by other pathogens - even genetically modified ones. There is also high incidence, particularly in children, of Rota virus infections. Flooding of the district of Malda in West Bengal due to heavy rain in July 1998 caused an epidemic of gastrointestinal infections. 42,000 persons, mainly women and children, were infected after drinking polluted water. Among the infected 381 died. Furthermore, it was reported that in the flood-hit districts of Bangladesh 571,863 people had been affected by waterborne disease, and 770 of them had died. An increase in morbidity and mortality would have to be very high to be observed and suspected as being caused by BTW agents. Sampling blood and tissues will present practical problems. National registration may be incomplete and the identity and age of people very uncertain. Lack of medical resources, food and clean water are the prevailing conditions. The strong influence of religion on society is important but seldom improves the situation. However, infections caused by BTW agents that are not identified and not properly treated could produce serious problems not only because of a large number of casualties. The BTW agent could be resistant to antibiotics and easily be distributed to other countries by contagious visitors from the affected area. In case of virus, slow and ineffective countermeasures can permit the virus to recombine to a deadly strain.

In order to ameliorate the situation, fundamental measures have to be taken such as boring of wells to obtain clean water and separate this from latrines and waste water. Education is certainly of great importance as a means to improve hygienic conditions. Furthermore, at least a rudimentary infrastructure to respond to such crises should be established. Clearly, outside help will be essential for such an undertaking.

6. Conclusions

We are living in a world with yawning chasms between developed countries with resources and poor countries that need resources and development. This includes the uneven distribution of clean water. Populations in developing countries already suffer from malnutrition and diseases. If they are expelled from their homes their situation is disastrous and conditions are created in which infectious diseases can spread. Despite local African epidemics of highly virulent viruses such as Ebola virus, it does not appear very probable that this strain of virus will become a widespread threat to mankind. Influenza pandemics caused by new combined super viruses could appear if the international network of surveillance fails. The situation for those infected by HIV, another virus changing its properties, could improve due to better medication. Unfortunately, there are reasons to fear that international terrorism could use BTW agents to achieve their goals even if in doing this they violate the BTWC.

7. References

1. Huxsoll, D.L. (1989) The impact of zoonoses on military operations, *Revue Internationale des Services de Santé de Forces Armées* **62 (10-12)**, 301-306.
2. Kiple, K.F. (ed.) (1997) *Plague, Pox & Pestilence*, Weidenfeld, London.
3. Crosby, W. (1989) *America's Forgotten Pandemic: The Influenza of 1918*, Cambridge University Press, Cambridge.
4. Kaplan, D.E. and Marshall, A. (1996) *The Cult at the End of the World. The Terrifying Story of Aum*, Crown Publishers Inc., New York.
5. Franz, D.R. et al. (1997) Clinical recognition and management of patients exposed to biological warfare agents, *JAMA* **278**, 399-411.
6. Larson, E. (1998) The Flu Hunters, *Time* **151 (11)**, 44-52.
7. DeJong, J.C. et al. (1997) A Pandemic Warning, *Nature* **389**, 554.
8. Shortridge, K.F. et al. (1998) Characterization of Avian H5N1 Influenza Viruses from Poultry in Hong Kong, *Virology* **252**, 331-342.
9. Taubenberger, J.K. et al. (1997) Initial Generic Characterization of the 1918 "Spanish" Influenza Virus, *Science* **275**, 1793-96.
10. Palese, P. and Roizman, B. (1996) Genetic engineering of viruses and of virus vectors, *Proc. Natl. Acad. Sci. USA* **93**, 11287.
11. Palese, P. et al. (1996) Negative-strand RNA viruses: Genetic Engineering, *Proc. Natl. Acad. Sci. USA* **93**, 11354-58.

Acknowledgements

I would like to thank Associate Professor Fredrik Elgh and Professor Anders Sjöstedt for stimulating discussions and good advice.

TABLE 1. Some data on Anthrax and Plague. Modified after reference [5]

Agent	Infective Dose (Aerosol)	Incubation Period	Diagnostic Samplers (BSL)*	Diagnostic Assay	Patient Isolation Precautions	Chemotherapy (Rx)	Chemo-prophylaxis (Px)	Vaccine Availability	Comments
Anthrax	8000 to 50 000 spores	1-5 d	Blood (BSL-2)	Gram stain Ag-ELISA, Serology; ELISA	Standard precautions	Ciprofloxacin 400 mg IV q 8-12 h Doxycycline 200 mg IV then 100 mg IV q 8-12 h Penicillin 2 million units IV q 2 h plus streptomycin 30 mg/kg IM qd (or gentamicin)	Ciprofloxacin 500 mg PO bidx4 wk. If unvaccinated, begin initial doses of vaccine Doxycycline 100 mg PO bidx4 wk plus vaccination	Michigan Biological Products vaccine (licensed): 0.5 mL SC at 0, 2, 4 wk and 6, 12, 18 mo, then annual boosters	Vaccine: boost at-risk annually Alternates for Rx: gentamicin, erythromycin, and chloramphenicol
Plague	100-500 organisms	2-3 d	Blood, sputum, lymph node aspirate (BSL-2/3)	Gram or Wright-Giemsa Stain Ag-ELISA, Culture, Serology; ELISA. IFA	Pneumonic: droplet precautions until patient treated for 3 d	Streptomycin 30 mg/kg IM qd in 2 divided doses x 10d (or gentamicin) Doxycycline 200 mg IV then 100 mg IV q 12x10-14 d Chloramphenicol 1 g IV q 6 hx10-14 d	Tetracycline 500 mg PO qidx7d Doxycycline 100 mg PO q 12 hx7d	Greer inactivated vaccine (licensed): 1.0 mL, then 0.2 mL, boost at 1-3 and 3-6 mo	Boost at-risk 12, 18 mo & yearly. Plague vaccine not protective against aerosol in animal studies. Alternate Rx: chloramphenicol or trimethoprim-sulfamethoxazole Rx: chloramphenicol for plague meningitis

TABLE 2. Parameters to be considered that affect the growing and dispersion of bacteria or viruses as aerosols or in water systems

	Aerosol Dispersion		Water
	Bacteria	**Virus**	**All types of microorganisms**
Microorganisms	Some problems to obtain. Must be purchased from strain collections or stolen, i.e. anthrax, pneumonic plague, tularaemia	Some or great problems to obtain. Must be purchased from strain collections or stolen, i.e. smallpox, hemorrhagic fevers, encephalitis	Easy to obtain. sludge, faeces, earth etc. can be used
Equipment Small-scale cultivation 10-200 litres	Easy to obtain: i.e. glass bottles, demijohns, aquatic pumps, immersion heater, cultivation media	Some problems to obtain i.e. cell cultivation bottles, cell lines, special warming cupboards	Easy to obtain, i.e. bath-tub
Large-scale cultivation > 200 litres	Hard to obtain; i.e. fermentors, equipment for control	Hard to obtain, special equipment for viruses	Easy to obtain: i.e. more bath-tubs

TABLE 3. Modified after [6]

FLU PANDEMICS: THEN AND NOW		
1918 What happened	Year	2000 What could happen
1.8 billion	World population	6.1 billion
Troop ships, railroad	Primary mode of transportation	Jets
4 months	Time for virus to circle the globe	4 days
Gauze, masks, disinfectants	Preventive measures	Vaccines
Bed rest, aspirin	Treatments	Some antiviral drugs
20+ million	Estimated dead	60 million?

THE POST EXPOSURE PROPHYLACTIC MEASURES AGAINST VIRAL BTW AGENTS

SERGEY V. NETESOV
State Research Center of Virology and Biotechnology "VECTOR"
633 159 Koltsovo, Novosibirsk Region
Russian Federation

1. Introduction

Viruses do not have their own energy supplying and structural-components synthesizing systems. That is the main difference between viruses and such pathogens as bacteria and rickettsia. Therefore they have to use intracellular structures and enzymes for their replication, so the critical stage of their reproduction is the stage at which they penetrate the cell. Once they have entered the cell viruses do not leave it until multiplying and fully maturating after which they can infect hundreds or even thousands of new cells. Bacteria do not really need cellular structures for their replication because their main requirements are energy sources and elementary structural components, such as amino acids, nucleic acids, carbohydrates, fatty acids or their components. The main, if not the only, way to defeat a virus is to attack it before it enters the living cell, whereas bacteria may be attacked throughout their life cycle taking place in a macroorganism. The time interval before a virus enters the eucaryotic cell or attacks the critical organ is sometimes prolonged. This allows the use of drugs or preparations for diminishing the symptoms or even for preventing the disease itself. One striking example has recently been sent for publication by researchers of our Center. The manuscript describes the preparation and use of hyperimmune serum for prophylaxis and therapy of Ebola virus infection [21]. In this chapter I would like to present some results from this publication, together with other data about preparations for post exposure prophylaxis of viral infections.

2. Research in the Russian Federation

After it was realized that Ebola virus posed a threat to public health (in 1976), it remained a great challenge to virologists. Although work has been conducted for thirty years, there are as yet no proven methods of prophylaxis and therapy

A. Kelle et al. (eds.), The Role of Biotechnology in Countering BTW Agents, 247–256.
© 2001 Kluwer Academic Publishers. Printed in the Netherlands.

of filovirus infections. Unfortunately, all attempts to develop effective vaccines against Ebola infection have been unsuccessful [1-5]. Thus, there is a real need for the development of other antiviral preparations, such as conventionally used in emergency cases occurring in laboratory workers. These are specific antiviral immunoglobulins.

The first data concerning the treatment of Ebola fever by immune human convalescent plasma were reported in 1977, [14] when infection of an investigator with Ebola virus in a laboratory and treatment procedures were described. However, it is well known that sources of human immune plasma are scarce and therefore the solution was to develop a type of antiviral preparation using animals as a source of immune plasma. At the end of the eighties we started to develop preparations using goats and sheep. First, we tested the possibilities of obtaining animal immune plasma by immunization with inactivated virus preparations. However, the ELISA and neutralization Ebola antibody specific titers we measured were unacceptably low, which was in agreement with the previously published data [2, 4-5]. Subsequently, we found that sheep and goats were completely insensitive to Ebola virus and that we could, consequently, use large quantities of live virus for immunization. We continued utilizing live virus from infected guinea pigs as 10% liver homogenate. Virus neutralizing activity was tested in virus neutralizing experiments on guinea pigs; a specially selected Ebola virus strain, pathogenic for guinea pigs, was used for this purpose. This strain, called 8ms, was first selected and characterized by researchers from our VECTOR center [5]. Only batches containing virus neutralizing antibodies with titers above 2.5 Log_{10} were used in gammaglobulin preparations. As a result of fractionation, the concentrations of virus neutralizing antibodies (VNA) were more than 3 Log_{10} in the final immunoglobulin preparations [7, 15].

Analysis of the anaphylactic activity of these preparations on guinea pigs demonstrated that equine immunoglobulin preparation was the most anaphylactogenic compared to sheep and the caprine immunoglobulins.

The results of tests of the storage properties of the final immunoglobulin preparations at 4°C demonstrated that at least goat preparations may be stored for more than five years without a significant reduction in their neutralizing activity (all the other parameters of these preparations remained unaltered).

We also analyzed the neutralizing activities of different fractions of goat immune serum (described in [7, 15]). The fractions were obtained by ion-exchange chromatography. These data are given in Table 1. As shown, the IgG-2 fraction contains the bulk of antiviral antibodies. The IgG1A and IgG1B fractions mainly contain antibodies against liver antigens of guinea pig (see details in [7]). A similar analysis of the final immunoglobulin preparations obtained using Cohn's method showed that they contain mainly the IgG-2 antibodies. Furthermore, IgG2 fraction was less reactive in anaphylaxis tests than the other two fractions.

Goat immunoglobulin preparation was tested on guinea pigs for its ability to protect against Ebola virus infection. Immunoglobulin was injected 24, 48 and

72 hours before challenge by intraperitoneal inoculation of Ebola virus 8mc strain, pathogenic for guinea pigs. The results are given in Table 2. From the data of this table it is clear that the prophylactic index is maximal when there is not more than a 24 hour interval between immunoglobulin injection and virus inoculation. Table 3 presents the results of similar experiments in which immunoglobulin injections were made after inoculation with virus. In these cases, the results were positive up to 72 hours after virus inoculation.

Goat preparations were tested in conformity with national regulations for preclinical trials (anaphylactogenicity, apyrogenicity, acute and chronic toxicity, tumorogenicity, effects on hematological parameters, liver function). Clinical trials on volunteers were started after formal permission was given. Goat immunoglobulin was injected into 7 volunteers (the treated group) and placebo (the untreated control group) to 3 volunteers. There were no local or general reactions at the time of the injection and after it. Recorded body temperature, blood pressure and urine parameters were normal.

The IgG level in serum on day 7 rose to 7.8 units in the treated group and remained at this level until day 28, thereafter it slowly declined to the initial mean. IgM level in serum also rose on day 7, it then fell during days 14-21. There were no such variations in serum IgG and IgM in the control group during the observation period.

Based on the obtained results, national permission was obtained to use this goat immunoglobulin as an emergency prophylaxis preparation against Ebola infection. The preparation was used against Ebola in four humans. In two cases, there was a slight suspicion of infection with Ebola virus: hands were injured whilst washing cages containing infected dead animals with bleach solution. In the third case, the left hand was injured by a syringe after taking a blood sample from a monkey infected with Ebola virus (Zaire strain) four days before. The syringe was treated with bleach for 1-2 minutes before the injury. In all the above cases, the immunoglobulin administration was combined with injections of recombinant alpha-2-interferon (two times daily for 14 days). The only symptom of illness was a rise in temperature to 37.3-37.5°C, probably because of the injected interferon.

The fourth accident with suspected Ebola infection happened to an investigator who pricked his left palm with a thick needle during a plasmapheresis procedure with an ill monkey earlier infected with Ebola virus (Zaire strain). An additional analysis of the blood from this monkey, taken on the day of the accident, demonstrated the presence of live virus at a concentration of 7.8 $Log_{10}LD_{50}$ (for guinea pigs). According to our calculations, the injured person received at least 10^2-10^3LD_{50} baboon dose. This fourth case is noteworthy because there were some other symptoms related to possible Ebola infection. To begin with, he tried to press out all the blood from the wound and treated it with bleach. Then he was referred to P4 hospital where he received injections of anti-Ebola goat immunoglobulin (6 ml intramuscularly). Recombinant human alpha-2-interferon was administered two times daily, in 3 mln. IU portions for 12 days. Laboratory tests of 20 biochemical parameters of the blood were performed daily. Four days after injury α-amylase level in serum increased twofold. On day 5, the size of the inguinal lymph nodes increased and so did that of the liver, by 2 cm. The rise in body temperature to 37.0 - 37.3°C recorded from the first day of injury was presumably due to interferon treatment. On day 6, fibrinogen level started to rise, reaching 5.5 g/l (its normal value does not exceed 4 g/l); the same was observed for the prothrombic index which rose to 112% (its normal percentage being 100%). The activities of aminotransferases ALAT and ASAT increased, too. There was erythema at the site of immunoglobulin injection. Symptoms of illness included myalgia, headache, arthralgia, malaise. On days 8 and 10, a hemosorption procedure was performed. All the parameters normalized on 12 day after the injury. The patient's general condition improved, the symptoms disappeared. Finally, he recovered. The virus was not detected, nor were antibodies found in his blood.

A retrospective analysis of his case report revealed a puzzling pattern. The course of changes in biochemical parameters showed that a cytolytic process of unknown etiology was developing. Clearly, all the symptoms could not be attributed to Ebola infection or be reactions to gammaglobulin injection. However, the volunteers showed no similar reactions in trials of the preparation.

Similar immunoglobulin technology development and protective efficacy studies were made in parallel and partly independently of ours at the Virological Center (located in Sergiev Possad near Moscow) of the Institute of Microbiology of the Russian Ministry of Defense. An immunoglobulin preparation from immunized horse serum was successfully developed at this Center [3, 16, 17]. The Moscow virologists used a similar live virus preparation to immunize horses (described in [11]) and a similar immunization schedule [3, 11, 16]. As a result, they developed immunoglobulin preparations not only with the titers 1:8000 but also as high as 1:64000 in VNA test (see [16]). According to their first results, the preparations protected baboons against low doses of Ebola virus (10-30 LD_{50} by intramuscular infection). The only fatally infected baboon died after a prolonged incubation period on day 18 postinfection, which is quite unusual for Ebola virus.

In their second and third papers [16, 17], the Moscow investigators gave a more detailed evaluation of their protective efficacy studies of equine gammaglobulin in baboons. The relevant information is given in Table 4 (see [16] for details). The preparation with the highest VNA titers obviously provided a more than 80% protection against Ebola virus injection (the injected virus dose was 10-30 LD_{50}); the infected baboons did not develop viremias in blood and liver only when immunoglobulin was injected during the first hour postinfection. When the interval between virus injection and immunoglobulin inoculation was more than 1.0-1.5 hours, the percentage of baboons surviving was considerably smaller. It should be noted that in this particular case, the incubation period was up to 26 days, much longer than usual (4-7 days). The immunoglobulin preparation met all the national standard requirements, and it was evaluated by the

State L.A.Tarassevich Institute of Standardization and Control of Medical Immunobiological Preparations of the Russian Federation. The product was formally approved for use in Russia for emergency treatment of Ebola fever from 1995.

3. Discussion

The United States Army Research Institute of Infectious Diseases (USAMRIID, Fort Detrick, U.S.A.) received doses of equine anti Ebola immunoglobulin and independently conducted protective efficacy studies [18]. The results obtained in cynomolgus monkeys treated with the immunoglobulin simultaneously with inoculation with Ebola virus were thought-provoking: the beneficial effects of IgG treatment were limited to a delay in the onset of viremia and clinical signs when compared to untreated controls. The discrepancy between these results and those reported by the Moscow investigators [16, 17] may be explained by the use of other monkey species and large (up to 1000 LD$_{50}$) infectious doses [18].

Specific gammaglobulin administered early after infection of guinea pigs and Hymadryl baboons is effective presumably because small quantities of Ebola virus are completely neutralized by antibodies *in vivo* before the virus starts to interact with cell receptors. It is well known that human hyperimmune anti tick-borne encephalitis (TBE) virus immunoglobulin is similarly effective against low doses of TBE and that this preparation has been successfully used for many years in Austria and Russia for urgent prophylaxis of disease in the case of tick bite. As for Ebola virus, its presence was first shown in monocytes and macrophages of infected monkeys 48 hours postinfection [19]. Specific antibodies against Ebola virus after immunoglobulin inoculation of baboons appear in the bloodstream as early as 30 min postinfection. They reach maximum concentration after 1 hour and continue to circulate for up to 10 days, decreasing in concentration. Neutralization of Ebola virus by immunoglobulin takes 1 hour *in vitro*. Taken together, all these facts may explain why the equine anti Ebola immunoglobulin preparation is effective when administered no more than 1-2 hours after Ebola virus infection.

Heterologous immunoglobulins are not used widely because of significant reactogenicity for humans and the fact that they cannot be used repeatedly because they cause anaphylactic shock. But sometimes there is no alternative as a preventive measure against diseases caused by such agents as Ebola or Marburg viruses, poisonous snake bites or other poisonous bites.

Homologous immunoglobulins are well-known antiviral preparations whose effectivity has been proven and multiply confirmed in different countries, both for treatment and for prevention of disease. The appropriate hyperimmune immunoglobulins injections are successfully used for post–exposure prevention in case of possible infection by hepatitis A virus, influenza virus, measles virus, and tick-borne encephalitis virus. The effect of these injections is maximal when they are used during the first 24-48 hours after the suspected moment of infection, and this may be explained by the length of time needed for the virus to enter the cell. In this case immunoglobulin molecules bind with the virus, marking it enough for recognition by immune cells of the macroorganism and for subsequent destruction, or even making it impossible for the virus to enter the cell.

Another example is the use of chemical compounds, which oppose the penetration of viruses into the cells and therefore block the infection process. Rimantadine (Russia) or amantadine (USA and Europe) or their analogues are the only known examples of drugs that are really effective in case of influenza virus infection as preventive or emergency post-exposure drugs.

The third and final method is made possible by the extremely slow process of replication of some viruses. The typical and practically the only example is rabies virus which can only be combated by post exposure vaccination by attenuated or inactivated viral preparation. This unique opportunity exists because the incubation period for this virus is more than thirty days, whereas the development of complete immune response takes 14 days only. This feature of rabies virus makes post exposure vaccination possible so preventing the infection process and the disease itself. There are only a few other viruses that have a similarly long incubation period: hepatitis A, HIV-1 and -2, and some others. I have found no data indicating that it is possible to prevent hepatitis A using this method, but am fairly sure that this should be possible. As for the HIV-1 and -2 viruses, similar methods of prevention may be possible but have not yet been experimentally proved.

4. Conclusion

Post exposure prevention measures exist in some viral infections because the viruses take a long time to enter the cells. There are three ways of doing this:
1. Using specific hyperimmune homologous or heterologous immunoglobulins (hepatitis A infection, Ebola infection, influenza, tick-borne encephalitis virus, venom toxins and some others);
2. Using chemical drugs which prevent the penetration of the virus into the cell (rimantadine in the case of influenza virus);
3. Post exposure vaccination (rabies virus and, possibly, hepatitis A virus).

The sources of donor plasma for the preparation of homologous immunoglobulins are very limited, whereas the heterologous preparations are reactogenic. Therefore, the alternative processes including recombinant human antibody technology should be developed and industrialized for this purpose.

5. References

1. Lange, G.V., McCormick, G.B., Walker, D.H., Kiley, M.P., and Neidhardt F.S. (1985) Vaccination of rhesus monkey by gamma-inactivated Ebola virus and results of live virus challenge in immune and native animals, *Abstr. Ann. Meet. Amer. Soc. Microbiol.,* Las Vegas, Nevada, 3-7 March, 1985; CDC, Atlanta; p. 295.
2. Lupton, H.W., Lambert, R.D., Bumgardner, D.L., Moe, J.B., and Eddy, G.A. (1980) Inactivated vaccine for Ebola virus efficacious in guinea pig model, *Lancet* **2(8207)**, 1294-1295.
3. Mikhailov, V.V., Borisevich, I.V., Chernikova, N.K., Potryvaeva, N.V., and Krasnianskii, V.P. (1994) The evaluation in hamadyas baboons of the possibility for the specific prevention of Ebola fever, *Voprosy Virusologii* (in Russian) **39(2)**, 82-84.
4. Ignatyev, G.M., Streltsova, M.A., Agafonov, A.P., and Kashentseva, E.A. (1995) Mechanisms of protective immune response in monkeys with Marburg virus, *Voprosy Virusologii* (in Russian) **40(3)**, 109-113.
5. Chepurnov, A.A., Chernukhin, I.V., Ternovoi, V.A., Kudoiarova, N.M., Makhova, N.M., Azaev, M.S., and Smolina, M.P. (1995) Attempts to develop a vaccine against Ebola fever, *Voprosy Virusologii* (in Russian) **40** **(6)**, 257-260.
6. Merzlikin, N.V., Chepurnov, A.A., Istomina, N.N., Ofitserov, V.I., and Vorobyeva, M.S. (1995) Development and use of enzyme immunoassay test systems for the diagnosis of Ebola fever, *Voprosy Virusologii* (in Russian) **40 (1)**, 31-35.
7. Dedkova, L.M., Kudoiarova, N.M., Chepurnov, A.A., and Ofitserov, V.I. (1994) Composition and immunochemical properties of goat immunoglobulins against the Ebola virus, *Voprosy Virusologii* (in Russian) **39 (5)**, 229-231.
8. Moe, J.B., Lambert, R.D., and Lupton, H.W. (1981) Plaque assay for Ebola virus, *J. Clin. Microbiol.* **13 (4)**, 791-793.
9. Cohn, E.J., Strong, L.E., and Hughes, W.L. (1946) Preparation and properties of serum and plasma proteins. IV. A System for the separation into fractions of the protein and lipoprotein components of biological tissues and fluids, *J. Amer. Chem. Soc.* **68 (3)**, 459-475.
10. Van der Groen, G., Jacob, W., and Pattyn, S.R. (1979) Ebola virus virulence for newborn mice, *J. Med. Virol.* **4 (3)**, 239-240.
11. Krasnianskii, V.P., Mikhailov, V.V., Borisevich, I.V., Gradoboev, V.N., Evseev, A.A., and Pshenichnov, V.A. (1994) Preparation of hyperimmune horse serum against Ebola virus, *Voprosy Virusologii* (in Russian) **39 (2)**, 91-92.
12. Ashmarin, I.P. and Vorob'ev, A.A. (1962) *Statistical methods in microbiological investigation,* Medicine, Leningrad, p. 230. (in Russian)

13. Burgasov, P.N. (ed.) (1978) *Handbook on vaccination and seroprophylaxis,* Medicine, Moscow, p. 439 (in Russian)
14. Emond, R.T.D., Evans, B., Bowen, E.T.W., and Lloyd, G.N. (1977) A case of Ebola virus infection, *Brit. Med. J.* **2 (6086)**, 541-544.
15. Dedkova, L.M., Kudoyarova, N.M., Shaprov, V.N., Sabirov, A.N., and Offitserov, V.I. (1993) Antibodies of hyperimmune sera of animals. II. Subclasses of Immunoglobulin G from normal and hyperimmune sera of goat, *Siberian Biological Journal* (in Russian) **6**, 8-13.
16. Borisevich, I.V., Mikhailov, V.V., Krasnianskii, V.P., Gradoboev, V.N., Lebedinskaia, Ye.V., Potryvaeva, N.V., and Timan'kova G.D. (1995) Development and study of the properties of immunoglobulin against Ebola fever, *Voprosy Virusologii* (in Russian) **40 (6)**, 270-273.
17. Markin, V.A., Mikhailov, V.V., Krasnianskii, V.P., Borisevich, I.V., and Firsova, I.V. (1997) Developing principles for emergency prevention and treatment of Ebola fever, *Voprosy Virusologii* (in Russian) **42 (1)**, 31-34.
18. Jahrling, P.B., Geisbert, J., Swearengen, J.R., Jaax, G.P., Lewis, T., Huggins, J.W., Schmidt, J.J., LeDuc, J.W., and Peters, C.J. (1996) Passive immunization of Ebola virus-infected cynomolgus monkeys with immunoglobulin from hyperimmune horses, *Archives of Virology – Supplementum* **11**, 135-140.
19. Ryabchikova, E.I., Kolesnikova, L.V., Smolina, M., Tkachev, V.V., Pereboeva, L.V., Baranova, S.B., Grazhdantseva, A.A., and Rassadkin, Yu.N. (1996) Ebola virus infection in guinea pigs: presumable role of granulomatosis inflammation in pathogenesis, *Archives of Virology* **141 (5)**, 909-922.
20. Fields, B.N., Knipe, D.M., Howley, P.M. (eds) (1996) *Fields Virology,* Raven Publishers, Philadelphia/New York.
21. Kudoyarova-Zubavichene, N.M., Sergeyev, N.N., Chepurnov, A.A., and Netesov, S.V. (1999) Preparation and Use of hyperimmune serum for prophylaxis and therapy of Ebola virus infection, *Journal of Infectious Diseases* **179**, Suppl 1, S218-223.

Acknowledgements

I would like to thank my colleagues Drs. Alexander A. Chepurnov, Natalia M. Kudoyarova-Zubavichene and Nikolai N. Sergeyev who carried out most of the experimental work, Dr. Larissa M. Dedkova who participated in fractionation and analysis of the sera and immunoglobulin preparations, Dr.Sergey V. Luchko who performed most of the neutralization experiments, and Dr. Margarita P. Smolina whose significant contribution to experimental selection of Ebola virus strain pathogenic for guinea pigs facilitated our work.

TABLE 1. Ebola virus neutralizing activity of fractions of goat immune serum

Fraction	Log_{10} neutralization index (LNI)			
	Experiment #1	Experiment #2	Experiment #3	Mean LNI ± SD
Immune serum	2.25	3.0	2.75	2.66±0.22
IgG 1A	2.25	2.0	1.5	1.92±0.22
IgG 1B	1.00	1.25	1.0	1.08±0.083
IgG 2	2.75	3.25	3.0	3.0±0.144

TABLE 2. The prophylactic effect of goat anti-Ebola immunoglobulin inoculation on guinea pigs before infection with Ebola virus related to the time interval

Experiment No	Log10 prophylaxis index (LPI) related to the time interval between immunoglobulin inoculation (before infection) and virus injection		
	72 hours	48 hours	24 hours
1	0.74	1.25	2.0
2	1.0	1.25	1.75
3	0.75	1.5	2.0
Mean LPI ± SD	0.83 ± 0.085	1.33 ± 0.085	1.92 ± 0.83

TABLE 3. The prophylactic effect of anti-Ebola immunoglobulin inoculation on guinea pigs related to the time interval after infection

Experiment No	Log_{10} prophylactic index (LPI) related to the time interval between virus infection and immunoglobulin inoculation after infection (hours)					
	4	24	48	72	96	120
1	0.75	1.75	1.25	0.5	0.5	0
2	1.0	1.75	1.25	0.5	0	0
3	1.25	2.5	0.75	0.75	1.0	0.25
Mean LPI ± SD	1.0±0.14	2.0±0.25	1.1±0.17	0.6±0.08	0.5±0.29	0.08±0.1

TABLE 4. Evaluation of the effectiveness of anti-Ebola virus gamma globulin in baboons (*Papio hamadryas*) infected by intramuscular injection of Ebola virus (presented as in [17])

Time of injection with respect to infection (immunoglobulin dose - 6 ml)	VNA* titer in immunoglobuli n preparation	Number of treated and percentage of survived baboons		Incubation period for ill baboons (days)	Longevity of ill baboons (days)	Baboons with viremias(%) (days postinfection)				
		taken into experiment	survived, %			5-6	7-8	10-11	13-15	20-33
2 hours before infection	1:8192	2	100	-	-	0	0	0	0	0
5-15 min.after inf.	1:4096	6	50	9±4	18±9	33	0	0	40	0
	1:65536	10	100	-	-	0	0	0	0	0
30 min.after inf.	1:4096	5	0	7±1	10±1	-	100	100	0	0
	1:65536	14	80	11	14±2	0	0	21	0	0
60 min.after inf.	1:4096	5	20	9±4	11±5	-	40	50	50	0
	1:16384	5	100	-	-	0	0	0	0	0
120 min.after inf.	1:4096	3	0	9±3	13±2	-	100	100	0	0
	1:16384	7	29	10±2	26±11	0	14	43	67	0
Control (infection with the same dose without gammaglobulin)		20	0	6±2	9±1	100	100	100	-	-

* - VNA - virus neutralizing antibodies.

SEARCH OF BIOTECHNOLOGY-BASED DECONTAMINANTS FOR C/BW AGENTS

R. DIERSTEIN, H.-U. GLAESER, A. RICHARDT
German Armed Forces Scientific Institute
for Protection Technologies - NBC-Protection
Humboldtstraße 1
D-29633 Munster
Germany

1. Introduction

Current decontamination procedures for C/BW agents depend on temperature treatment and chemically reactive components, e. g. strong oxidants and chemical disinfectants. However, these techniques and compounds are not always suitable for different materials. Furthermore, they cause logistic problems and are themselves hazardous, corrosive and questionable from an environmental viewpoint. In the German Federal Armed Forces hypochlorite and formaldehyde are presently the main reactive components for C/BW-decontamination. The common goals of the various treatments are to rapidly destroy known CW agents as well as to disinfect bacteria, fungi and protozoa, inactivate viruses and detoxify toxins. For all decontamination procedures applied to clothing, material and vehicles the German Armed Forces prefer a combined technique for CW and BW. The decontamination of BW agents including bacterial spores and low molecular mass toxins, which have proved to be highly resistant to temperature and disinfectants, has become a particular challenge with respect to optimizing the parameters of efficiency, logistic burden and environmental impact. In order to search for novel reactive components we have started to focus on biotechnology-based decontaminants. As will be described in this paper, several biotechnological approaches may become appropriate alternatives to existing protective measures against C/BW agents. Moreover, they might in many cases be superior to the latter, as they allow reactions under mild conditions and follow mainly a highly effective catalytic rather than stoichiometric principle.

A. Kelle et al. (eds.), The Role of Biotechnology in Countering BTW Agents, 257–265.
© 2001 *Kluwer Academic Publishers. Printed in the Netherlands.*

2. Results and Discussion

2.1 RECOMBINANT ENZYME FOR CW-DEGRADATION

One of our most relevant recent developments in B/C-decontamination was a successful study of enzymes which can rapidly decompose nerve agents and - after a chemical modification - blister agents. We started a decontamination program on biotechnology-based reactive components with the recombinant diisopropylfluorophosphatase (DFPase, EC 3.1.8.2) from the head ganglion of *Loligo vulgaris* (squid), which was previously cloned and sequenced in an external research project at the University of Frankfurt, Germany (patented). This enzyme has a size of 35 kDa and is comprised of a single chain of 314 amino acids containing 8 cystein residues. It is stable at temperatures of up to 55 °C and has a pH-stability between 5.5 and 10 with a maximum activity at a near neutral pH of 7.5. It has a high specific activity of 200 units corresponding to 200 μmol diisopropyl fluorophosphate/mg enzyme/minute.

Laboratory tests with this enzyme were performed at Munster using different CW-agents of the G-type, soman, sarin, cyclosarin and tabun in a simple and environmentally compatible buffer of ammonium hydrogencarbonate at a pH of 7.5. The results and the experimental procedures are shown in Fig 1. As a control of spontaneous hydrolysis in water, the same experimental conditions were used in the absence of enzyme. At a given enzyme concentration of 30 units per ml and an agent concentration of 200 μg per ml, over 95% of each agent was degraded during the first hour. Sarin showed a complete degradation even after 20 minutes.

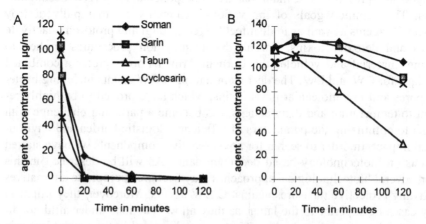

Figure 1. Degradation of an aqueous CW-agent mixture by recombinant diisopropyl-fluorophosphatase. Samples were incubated at 20 °C at a concentration of 200μg/ml in the (A) presence and (B) absence of enzyme (30 units/ml), 50 mM ammonium hydrogen-carbonate and 1 mM calcium chloride. Samples were taken at various times of incubation and subjected to standard gas chromatographic analysis.

Based on these positive results with a mixture of G-agents, which cover in total almost 80% of the spectrum, the use of hydrolytic enzymes for the decontamination of HD (mustard) was also investigated. In order to transform HD into a suitable enzyme substrate it was derivatized prior to the enzymatic treatment. One approach was the use of several nucleophilic phosphothiolate salts to form HD-organophosphate chimers (experiments performed in an external study at the Israel Institute for Biological Research, Ness-Ziona, Israel). As the HD derivatization could be achieved very rapidly with a half reaction time of only several seconds, the application of the HD-organophosphate chimer approach to a practical enzymatic decontamination solution for surfaces, and especially for skin, seems quite promising.

2.2 ENZYMATIC BW-DECONTAMINATION

Current decontamination strategies for equipment, clothing and vehicles contaminated with BW agents are mainly based on procedures which include heat and chemical-physical treatment. Table 1 summarizes the different techniques in comparison to possible biotechnological approaches. Hot air heated up to over 160°C is used for a combined decontamination of CW and BW agents, especially hot gas containing water and disinfectants. Denaturation is performed using disinfectants with formaldehyde, alcohols, organic and inorganic acids and detergents. These classical methods are used for decontaminating surfaces and skin, whereas an aerosolized form of the substances is used for various interiors. There are a number of additional physico-chemical techniques such as ozonization, which was tested with positive results for its use in water disinfection and purification. Other candidates were sonochemistry, a method of peroxide generation from water by sonication with very high ultrasonic frequencies in the range of 200 to 1200 kHz, and photolysis for irradiation of water with strong UV light. The latter was mainly effective against CW agents and toxins.

TABLE 1. Strategies for BW-Decontamination

Method	Principle	Specification
Conventional	Thermal	Hot air
		Hot gas
	Chemical-physical	Denaturation
		Ozonization
		Sonochemistry
		Photolysis
Biotechnical	Enzymatic	Muramidases
		Proteases
		Nucleases

All existing methods used so far are effective, but suffer primarily from the fact that they have to be conducted under high temperatures, intensive irradiation and/or with relatively high concentrations of hazardous, reactive components. As an alternative method for attacking bacteria, viruses and toxins, we started experiments using biocatalysts, which interact specifically with biological agents through catalytic processes under mild conditions. From literature studies we selected candidate enzymes such as: muramidases and lysozyme, which attack and degrade the bacterial cell wall leading ultimately to breakage of the cells; proteinases, which attack bacterial surfaces, viruses, protein and peptide toxins and degrade structural and functional proteinaceous components of BW-agents; nucleases, which destroy the genetic information contained in the desoxyribonucleic acid (DNA) and the ribonucleic acid (RNA) of viruses and bacteria.

Most critical for achieving decontamination or disinfection are the thermoresistant bacterial spores and toxins. It emerges that formaldehyde is still the most effective sporicidal disinfectant for surfaces, protection masks, vehicles and rooms. At present a combination of hot air gas with disinfectant injection (0.6 % of a commercial disinfectant containing formaldehyde and glutardialdehyde at a total concentration of over 10% (w/w)) has been tested and approved for B-decontamination in the Federal Armed Forces. In this regard, 450 liters of a 0.6 % solution of disinfectant containing formaldehyde are necessary to successfully decontaminate thermoresistant spores in a prototype system for heavy equipment, e. g. a tank. To remove the chemicals after a treatment of about 10 minutes duration, a washing step with high pressure and hot air must be carried out. This demonstrates that the existing technique imposes a considerable logistic burden on operations that might be lowered by employing alternative technologies or a suitable combination with novel techniques. So far, no enzymatic treatment has been developed to digest bacterial spores on a technical scale; however, we will also address this problem in our laboratory studies.

During recent years we have tested several protection measures against toxins. Various methods were applied to decontaminate surfaces and water from mycotoxins and especially the heat resistant algal toxin microcystin as model toxins in our laboratories. Table 2 depicts the different deontamination measures. The main principles were either to simply remove the toxins, to actually detoxify them or a combination of the two approaches. For elimination of the low molecular mass peptide microcystin from water, reverse osmosis was tested and proved to be very effective. Over 99% of the toxin was eliminated in a comparative study with commercially available membranes in laboratory scale devices. The exact retention value has yet to be determined, as the accuracy of the analysis was dependent on the detection limit of the analytical systems used earlier. These have now been improved with the introduction of new immunochemical and biochemical methods having detection limits that are three orders of magnitude lower than they were in previous systems.

TABLE 2. Protection Measures Against Toxins

Principle	Technique
Elimination	
Washing	Warm water/emulsion
Adsorption	Charcoal/resins
Precipitation	Iron chloride
Retention	Reverse osmosis
Detoxification	
Oxidation	Hypochlorite/peroxide
Photochemistry	UV-irradiation
Sonochemistry	Ultra sonication
Combination	UV-ozone
Biocatalysis	Enzyme reaction

In order to detoxify water, different physico-chemical methods were examined. The most effective of these approaches was the combination of ozone treatment and UV irradiation. However, high concentrations of ozone and intensities of UV-light irradiation were necessary. As depicted in Table 3, the model toxin microcystin was highly resistant to temperature, acids and a variety of proteases and peptidases. Biotechnological approaches for detoxification of water seemed to be impossible until 1996, when Bourne *et al.* [1] first published the enzymatic pathway of microcystinase, an enzyme with a biocatalytic activity that can degrade microcystin naturally.

TABLE 3. Resistance of the Heptapeptide Toxin
Microcystin to Degradation

Event	Specification	Stability
Heat	< 218 °C	+
Acid	pH 1	+
Digest	Trypsin	+
	Chymotrypsin	+
	Proteinase K	+
	Thermolysin	+
	Subtilisin	+
	Staphylococcus V8 Protease	+
	Serratia Proteasen	+
	E. coli leader peptidase	+
	Horseradish peroxidase	+

Table 3 lists the results of the stability studies with microcystin-LR. Aqueous solutions of the toxin at concentrations of 10–100 µg/ml were exposed for 90 minutes either in sealed ampoules to temperatures up to 220 °C, or at 37 °C in an aqueous solution of 0.1 N HCl, or in a 10 mM Tris buffer with different hydrolyzing and oxidizing enzymes at a final concentration of 0.1 mg enzyme/ml, or the equivalent in units. After incubation, samples were subjected to quantitative chromatographic analysis following the method of Dierstein et al. [2].

The mode of action of microcystinase from a *Sphingomonas* species is sketched in Figure 2. The cyclic toxic heptapeptide microcystin-LR is cleaved at the peptide bond between the unusual amino acid adda ($2S,3S,8S,9S$-3-amino-9-methoxy-2,6,8-trimethyl-10-phenyldeca-$4E,6E$-dienoic acid) and the variable amino acid L-arginine. This cleavage leads to a linearized peptide which is no longer resistant to common proteases of the intestine. With the detection of this enzyme, combinations of biotechnical and physical treatments become possible. The most promising appears to be the combination of enzymatic digestion and reverse osmosis to effectively reduce the microcystin content of naturally or intentionally contaminated surface water, which might in extreme cases be the sole remaining source of drinking water for the troops. This is all the more important since the toxin is known to exhibit, in addition to its high lethal toxicity (LD_{50}=50µg/kg, intraperitoneally for mice), a tumor promoting activity at much lower concentrations. A proposed concentration limit of as little as 0.01 µg of microcystin per liter of drinking water is being discussed internationally as a measure to avoid chronic intoxication with an enhanced risk of liver cancer.

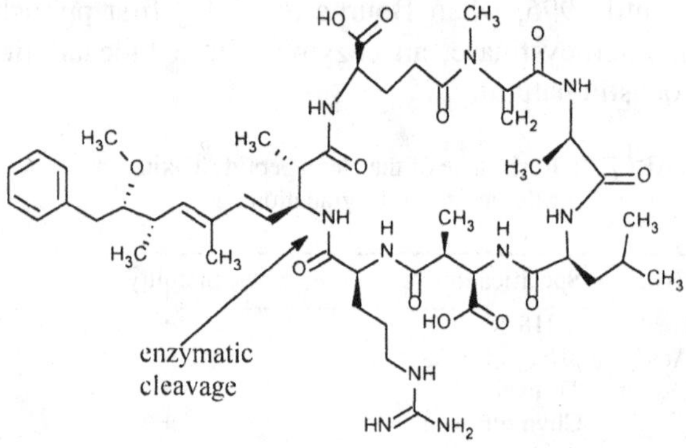

Figure 2. Microcystin-LR, a very heat and acid stable cyclic peptide toxin, is resistant to known proteases and peptidases. The unique cleavage site by microcystinase from *Sphingomonas* is indicated by the arrow. After data from Bourne et al.[1] and Rinehart et al. [3].

2.3 BW-DECONTAMINATION WITH BIOGENIC DISINFECTANTS

Besides enzymes, several biogenic antimicrobial and disinfectant agents have considerable potential for a biotechnical approach to BW-decontamination. One of the candidate molecules is protamine, a basic peptide (pI >10) with a length of 32 amino acids of which 21 are arginine [4]. It has microbicidal activity against a broad spectrum of bacteria and fungi, which is considered to be a -

result of its polycationic nature. The broad antibacterial spectrum and the fact that protamines are naturally occurring and non-toxic to humans make it a promising biotechnical alternative to chemical disinfectants. We therefore started a literature and experimental study to test the potential of protamines for the disinfection of BW-agents, including biotechnological production options for heterologous expression of the respective natural and synthetic genes. Based on the literature data analyzed so far, the protamine microbicidal action is more generic in comparison to e. g. bacteriocins, as the polycationic peptides most likely interact with anionic cell wall components and anionic phospholipids in the cytoplasmic membrane, thereby inducing condensation resulting in disruption of the cell envelope layers, killing both gram-positive and gram-negative bacteria. Recently, more specific data have become known for several bacterial species such as *Escherichia coli*, *Listeria monocytogenes* and *Shewanella putrefaciens* [5]. However, comparative experimental data with presumptive bacterial BW-agents are not yet available. We are presently concentrating on tests of combinations of protamine and enzymes as biogenic reactive components.

2.4 COMBINATION OF TECHNIQUES

As already discussed above, biotechnology-based decontaminants may be exploited for combinations of novel and existing techniques. In order to cover a broad spectrum of chemical agents, toxins, bacteria and viruses in one technical solution, a single approach is not likely to be sufficient or realizable. Therefore, our aim is to combine classical methods of decontamination with enzymatic components and to implant, in a first step, enzyme technology into existing systems of application. One of the experimental approaches to achieve this goal was the testing of the half-life of CW-degrading enzymes in the presence of conventional chemically reactive components. Figure 3 summarizes the results of this analysis (performed in an external study at University of Pittsburgh, USA). When organophosphohydrolase was exposed to even minute concentrations of hypochlorite it was rapidly inactivated. After a modification involving mixing the enzyme with polyethylene glycol, which is one method of stabilizing enzymes, an increase in resistance to the oxidizing chemical was observed. When the enzyme was immobilized in polyurethane a significantly high stability was achieved, even at increased concentrations of hypochlorite, which would be sufficient to exert an additional decontamination effect. We also started to test enzyme stability in the presence of alcohols and organic solvents. The results achieved so far demonstrate that immobilization is a promising technique for the integration of enzyme technology into existing and novel decontamination procedures by either replacing conventional hazardous reactive components or effectively reducing their concentration.

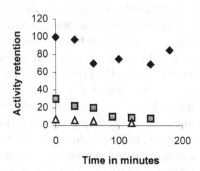

Figure 3. Enzyme stability in the presence of hypochlorite bleach. Organophosphohydrolase from *Pseudomonas diminuta* was incubated for various periods in an aqueous solution of (▲) native enzyme in 0.00625 wt% bleach; (■) polyethylenglycol modified enzyme in 0.00625 wt% bleach; and (◆), polyurethane immobilized enzyme in 0.1 wt% bleach

The experimental data available so far are very promising and there is good reason to assume that in about 5 years from now decontamination solutions or emulsions might include enzymes and biogenic biocides as reactive components. Suitable applications can be conventional sprayers for surface decontamination and aerosol disinfection. One recent article dealing with the integration of enzymes into existing techniques described the use of a combination of enzymatic decontamination with fire fighting foams as a carrier [6]. Our institute also very recently completed the development of an environmentally friendly fire fighting ECO-foam based solely on biodegradable components. We will also test this novel product as a carrier for enzymatic decontamination. The degrading enzymes can be directly added as a powder to the hose systems during operation at very low amounts: for CW-degrading enzymes, as little as 30 g of enzyme will be sufficient to produce 150 cm³ of detoxifying foam.

3. References

1. Bourne, D.G., Jones, G.J., Blakeley, R.L., Jones, A., Negri, A.P., and Riddles, P. (1996) Enzymatic pathway for the bacterial degradation of the cyanobacterial cyclic peptide toxin microcystin LR, *Appl. Environ. Microbiol.* **62**, 4086-4094.
2. Dierstein, R., Kaiser, I., and Weckesser, J. (1988) Rapid determination of *Microcystis sp.* toxins by reversed phase liquid chromatography, *FEMS Microbiol. Lett.* **49**, 143-147.

3. Rinehart, K.L., Namikoshi, M., and Choi, B.W. (1994) Structure and bio-synthesis of toxins from blue-green algae (cyanobacteria), *J. Appl. Phycol.* **6**, 159-176.

4. Ando, T.M., Yamasaki, M., and Suzuki, K. (1973) Protamines; isolation, characterization, structure and function, in A. Kleinzeller and H. G. Wittmann (eds.), *Molecular Biology, Biochemistry and Biophysics*, Chapman & Hall Ltd., London, pp. 1-109.

5. Johansen, C., Verheul, A., Gram, L., Gill, T., and Abee, T. (1997) Protamine-induced permeabilization of cell envelopes of gram-positive and gram-negative bacteria, *Appl. Environ. Microbiol.* **63**, 1155-1159.

6. LeJeune, K., Wild, J.R., and Russell, A.J. (1998) Nerve agents degraded by enzymatic foams, *Nature* **395**, 27-28.

BIOTECHNOLOGY AND POLITICO-MILITARY RESPONSES TO BTW AGENTS

MALCOLM R. DANDO CBiol FIBiol
Professor of International Security
Department of Peace Studies
University of Bradford
Bradford BD7 1DP
West Yorkshire UK

1.   Introduction

This volume began with a consideration of the wider political issues – particularly the strengthening of the Biological and Toxin Weapons Convention (BTWC) – within which the technical issues discussed in the following chapters are set. It would be standard practice, in these later chapters, to return to the wider political issues in the light of our technical discussions. However, that is not as straightforward here as we might normally expect - though modern biotechnology obviously has important benefits in countering BW agents, it is obviously also partly the cause of the increasing probability of their proliferation and use.

In this chapter, therefore, I begin by looking briefly at the nature of the threat and the range of means we have to counter that threat. This leads on to a consideration of the biotechnology revolution which I see as the latest of a series of scientific/technological developments which have had an impact on the evolution of biological weapons programs over the last century. My analysis of politico-military responses is then set within that historical framework.

2.   The Threat

In the early 1990s, in the view of the US military, there were four Concepts of Use (COU) that needed to be considered in regard to biological weapons: superpower *vs* superpower; state *vs* state; state *vs* factional element or *vice versa*; and terrorist use. [1] These military COUs are set out in Table 1.

A. Kelle et al. (eds.), The Role of Biotechnology in Countering BTW Agents, 267–283.
© *2001 Kluwer Academic Publishers. Printed in the Netherlands.*

TABLE 1. Military concepts of use of biological weapons *

Superpower vs Superpower

"During periods of the cold war ...the offensive programs of the two countries were orientated principally towards use on a scale commensurate with this perceived threat. Programs on this scale would be expected to be highly sophisticated, difficult to disguise completely...and to require extensive facilities... In addition to including smaller-scale weapons and simple delivery systems, a superpower BW program could include munitions delivery systems that are highly efficient, sophisticated, and provide large area coverage capability."

State vs State

"In a conflict between two less technologically advanced countries ... BW weapons might be desired and used by one country against another. In this COU, the quantities of agents needed for a limited number of weapons are likely to be far less than in a superpower program, and the types of delivery and dissemination equipment could be far less sophisticated.... The BW weapons are likely to be less efficient, more modest in sophistication and provide limited area coverage."

State vs Factional Element or vice versa

"In this COU, e.g., Iraq *vs* the Kurds, the targets are more limited, the quantities of agent needed are likely to be far less and the types of delivery and dissemination equipment could approach the primitive, yet still accomplish the goal..."

Terrorist Use

"Terrorist use of BW agents could be as simple as a knowledgeable individual with a grudge, or as complex as State-sponsored terrorism. The quantities of agent could be small ... production and purification methods extremely simple, and dissemination means simple to complex..."

---

* From reference 1

Today, a superpower *vs* superpower confrontation seems remote, but the other three possibilities are widely regarded as problems we have to face.

Wheelis has attempted to dissect the dimensions of such biological warfare by distinguishing between the nature of the aggressor, the scale of release of agent and the target of attack. [2] The categories within each of these dimensions are shown in Table 2.

TABLE 2. Dimensions of biological warfare *

1. Nature of the aggressor	2. Scale of release	3. Target
a. Nations	a. Point source release	a. Human
b. Subnational groups	b. Medium-scale release	b. Plants
c. Individuals	c. Large-scale release	c. Animals

* From reference 2

For each of the potential types of target – human, animal or plant – it is thus possible to define possible types of biological warfare by combining the nature of the aggressor with the scale of release of agent within a simple three-by-three matrix (Table 3).

TABLE 3. Types of biological warfare *

Scale of release	Nature of aggressor		
	Individual	Subnational group	State
Point source	e.g. criminal act	e.g. assassination	e.g. assassination
Medium scale	e.g. criminal act	e.g. terrorist	e.g. military tactical
Large scale	not possible	e.g. national liberation (army) use	e.g. military strategic

* From reference 2

It is obvious from this brief analysis that we face a multifaceted threat from biological weapons. As Miller [3] expressed it, in the official account of the participation of the US Air Force in the early years of the US offensive biological warfare program, "BW offers an endless variety of ways to wage war".

3.   The Web of Deterrence

With such a range of threats, it is not surprising that thoughtful analysts have argued for a wide range of policies to prevent the use of biological weapons. [4, 5, 6] Pearson has suggested that an integrated set of policies can be conceived of as a 'web of deterrence' able to convince a potential possessor that a biological weapons capability is not a worthwhile investment. [7]

The elements of the web of deterrence are set out in Table 4. The envisaged system consists of comprehensive arms control agreements, broad export monitoring and control, effective defensive and protective measures, and a range of determined national and international responses to the acquisition or use of biological weapons.

TABLE 4. Elements of the web of deterrence *

"– Comprehensive, verifiable, and global CB arms control to create a risk of detection and a climate of political unacceptability for CB weapons;

– Broad export monitoring and controls to make it difficult and expensive for a proliferator to obtain necessary materials;

– Effective CB defensive and protective measures to reduce the military utility of CB weapons; and

– A range of determined and effective national and international responses to CB acquisition and/or use."

* From reference 7

Other contributions have been centered on a study of the role of modern biotechnology in agent detection and protection, [8] but the main task of this chapter is to widen the discussion to incorporate some of the other elements of the policies we require to prevent biological warfare.

4. The Biotechnology Revolution

During recent years I have made a number of assessments of the impact of modern biotechnology on the evolution of future biological weapons. [9, 10, 11, 12] I do not intend to cover that ground again here, but several important points need to be noted. First, it would appear that as scientific understanding of microbiology – and the pathogenicity of infectious organisms – has grown since the pioneering discoveries of Pasteur, Koch and their colleagues at the end of the nineteenth century, this knowledge has been applied in the development of progressively more sophisticated offensive biological warfare programs. [13] This process is set out diagrammatically in Table 5. The 'Golden Age' of microbiology at the end of the nineteenth century, when the bacterial agents which caused many serious diseases (such as anthrax) were discovered and characterized, was followed by the sabotage campaigns against animal stocks by both sides in World War I. German use of anthrax was, of course, recently confirmed using modern analytical methods [14] on material captured in Norway some 80 years ago!

TABLE 5. Scientific advances and military developments *

CIVIL	MILITARY
'Golden Age' of Bacteriology	World War I/ Inter-War Years Programs
Aerobiology/ Industrial Production Capability	World War II/ Mid-Century Programs
'Golden Age' of Virology/ Genetic Engineering	Late Cold War/ Immediate Post-Cold War Programs
Genomics/ Molecular Medicine and Beyond	'Tailored' Classical Agents through to Novel Agents

* From reference 13

After the Great War ended, a number of countries, France for example, began careful scientific studies of the possible use of biological weapons. This process culminated in the massive Japanese offensive biological weapons program of the inter-war years and World War II. However, the Japanese do not appear to have given the attention necessary to the growing science of aerobiology, which informed the British and then the US offensive programs. Increasingly also, following World War II, the US program benefited from the growing industrial production capabilities which flowed from the need to overcome the difficulties of producing the new antibiotics.

It was only towards the end of these mid-century programs that the 'Golden Age' of virology in the 1950s facilitated the easier use of these agents, and then only in the 1970s and 1980s that genetic engineering became available. Not surprisingly, therefore, the Soviet Union's offensive biological weapons program was characterized by a greater concentration on viral agents and the use of genetic engineering to tailor agents in various ways for particular purposes.

Current US official estimates [15] suggest that although the 'classical' anti-personnel agents remain the greatest threat at present, because they have already been tested, the further tailoring of agents for specific purposes is to be expected in offensive programs in the near future (Table 6).

TABLE 6. Novel agents that could be produced by genetic engineering *

"– Benign microorganisms, genetically altered to produce a toxin, venom, or bioregulator.

– Microorganisms resistant to antibiotics, standard vaccines, and therapeutics.

– Microorganisms with enhanced aerosol and environmental stability.

– Immunologically-altered microorganisms able to defeat standard identification, detection, and diagnostic methods.

– Combinations of the above four types with improved delivery systems."

* From reference 15

If this argument is broadly correct, it is surely clear that, unless something significant is done to prevent it, the process will continue unabated. We shall see not only the beneficial improvements in detection and protection capabilities considered in this volume, but also continuing evolution of biological warfare agents.

In considering the implications of that statement, it is important to grasp the profound proportions of the biotechnology revolution, a revolution which is likely to shape our way of life at least as dramatically as the information technology revolution has done over the last few decades. [10] It is not just that we have the convergence of a variety of interacting scientific disciplines – what E. O. Wilson [16] has termed 'consilience' – which drives discoveries forward at an enhanced pace, but that these discoveries are leading to practical products from which industry can make a great deal of money. As a recent review in *Science* noted: [17]

> "...this new science of genomics is forcing some of the world's largest companies to reinvent themselves as borders between pharmaceutical, biotech, agricultural, food, chemical, cosmetics, environment, energy, and computer industries blur and erode..."

It added:

> "...Genomics has substantial government support, massive corporate investment, powerful enabling technologies, and short-term cash-generating potential."

It is, of course, very difficult to decipher where this process will lead us and what the implications will be for future developments of biological weapons if civilian innovations continue to be applied in the military sphere. There are, nevertheless, obvious ominous possibilities. Some of the underlying trends, as seen by the US military, [15] are set out in Table 7.

TABLE 7. Significant trends related to future biological weapons possibilities *

"– Genetically engineered vectors in the form of modified infectious organisms will be increasingly employed as tools in medicine and the techniques will become more widely available.

– Strides will be made in understanding of infectious disease mechanisms and in microbial genetics that are responsible for disease processes.

– An increased understanding of the human immune system function and disease mechanisms will shed light on the circumstances that cause individual susceptibility to infectious disease.

– Vaccines and antidotes will be improved over the long term, perhaps to the point where classical biological warfare agents will offer less utility as a means of causing casualties."

---

* From reference 15

We should be under no illusions about the difficulty of bringing the military consequences of such profound scientific revolutions under control. The potential dangers of the new science of chemistry were clear at The Hague Conferences at the turn of the last century, yet despite public revulsion at the large-scale use of chemical weapons in World War I, it was only at the end of the 20th century that an effective Chemical Weapons Convention came into force. [18] The first use of nuclear weapons was over half a century ago, but despite the considerable efforts made to maintain the nuclear non-proliferation regime centered on the Nuclear Non-Proliferation Treaty (NPT), there is little likelihood of eliminating such weapons with a Nuclear Weapons Convention over the next several decades. Chemical and nuclear technologies, however, are *old* technologies. Whilst biotechnology has long roots, genetic engineering, genomics and molecular medicine are *brand-new* technologies. The long process of bringing these technologies under proper regulation, even in the civil sector, has hardly begun. [19]

5. Politico-Military Responses

What then might be the best politico-military response to this set of circumstances? I would like to consider three sets of options: arms control; sanctions; and the use of force. These are clearly not exclusive options, but I will try to evaluate each in turn before looking at how they might fit together in the longer term. If we return to the idea of a web of deterrence to prevent the proliferation or use of biological weapons (Table 4), it will be noted that I begin by considering the first element of the web – the central norm that is embodied in the multilaterally agreed 1925 Geneva Protocol and the BTWC. As the creation and maintenance of the norm of non-use of biological weapons is the subject of the final chapter, [20] I shall deal directly only with the arms control aspect of the subject. I shall then move on to the fourth element of the web: international responses to transgression of the norm. I shall discuss two types of response: a range of sanctions; and the use of military force.

5.1 ARMS CONTROL

In the early 1990s, during the euphoria that accompanied the ending of the long east-west Cold War, there was a short period when it seemed to some that the need for arms control and disarmament agreements was over. [21] Now, almost a decade later, it can be seen that that view was incorrect. In point of fact, it arose from a misunderstanding about the nature of arms control.

Many people's understanding of arms control dates from the early 1960s, when a particular formulation was developed in the United States to deal with the problems considered important during the Cold War: [22]

> "...preserving a stigma against nuclear use through maintenance of a high threshold between conventional and nuclear weapons; the creation and preservation of redundant and survivable nuclear systems in the Soviet Union and the United States; and prevention of the spread of nuclear weapons to other states."

From this conception of arms control flowed the classic Cold War agreements: the Hot Line Agreement of 1963; the Partial Test Ban Treaty of 1963; The NPT of 1968; the Interim Agreement on Limitation of Strategic Offensive Arms (SALT I) of 1972; and the Treaty on the Limitation of Anti-Ballistic Missile Systems (ABM) of 1972. [23] But arms control in this formulation is primarily a bilateral exercise between two superpowers concentrated on nuclear systems and concerned overwhelmingly with arms race and crisis stability. Small wonder then that there appeared to be little place for such arms control in the early 1990s.

Yet as Joseph Goldblat has argued, in his major summary, arms control has actually come to include a very wide range of measures such as to: [24]

"(a) freeze, limit, reduce or abolish certain categories of weapons;
(b) prevent certain military activities;
(c) regulate the deployment of armed forces;
(d) proscribe transfers of some militarily important items;
(e) reduce the risk of accidental war;
(f) constrain or prohibit the use of certain weapons or methods of war; and
(g) build up confidence among states through greater openness in military matters."

Needless to say, the disjunction between the common conception of arms control and the actual problems and capabilities of arms control in practice led to many efforts to reformulate of the concept as the post-Cold War period began. [25, 26, 27] One of the most imaginative efforts was Stuart Croft's *History and Typology of Arms Control*. [28] Croft noted that:

> "Arms control is often seen to be a modern invention, a creation of the cold war. However, the practice of arms control is many thousands of years old..."

Essentially, he argued that over time political entities have developed a set of tools (political instruments) to deal with problems related to the regulation of military forces and that when the Cold War ended the international community had five main types of arms control at its disposal (Table 8). Clearly, arms control as generally understood during the Cold War is mainly contained within one sub-section of section 2 (strategic stability) of this typology. As the international community struggles to control the new forms of warfare characterizing the post-Cold War period [29], such a narrow conception of arms control is clearly inadequate.

TABLE 8. Types of arms control *

(1) Arms control at the conclusion of conflicts;
(2) Arms control to further strategic stability;
(3) Arms control to create norms of behavior;
(4) Arms control to manage the proliferation of weapons; and
(5) Arms control by international organization.

* From reference 28

The problems, of course, are not just caused by an inadequate conception of arms control. As Croft's historical approach makes clear, different forms of arms control have been found necessary in different historical periods. To find appropriate forms of arms control thus requires an assessment of the characteristics of the new international system. Bertrand, in a thoughtful paper, "The Difficult Transformation from 'Arms Control' into a 'World Security System'", has asked some of the necessary questions. [30] These questions can be seen in Table 9.

TABLE 9. Questions about the new international system *

(1) What will be the dominant features of the new system;
(2) What kinds of threats will be important;
(3) What means might best be used to address such threats
    – Types and sizes of armed forces
    – Measures of arms control;
(4) What processes and institutions might be most effective in reaching the necessary agreements within the system.

* From reference 30

Brad Roberts [31] has recently taken up this issue in more detail by asking what role arms control might have in helping to achieve six "world order tasks" which are defined as "steps useful for sustaining generally peaceful and orderly political-military relations within the interstate system". The six tasks are shown in Table 10.

TABLE 10. World order tasks *

(1) Maintaining stable relations among the major powers.
(2) Integrating aspiring powers.
(3) Insulating the state system from localized conflicts.
(4) Deterring aggressors and punishing transgressors.
(5) Managing technology diffusion.
(6) Engaging the United States in its 'unipolar' moment.

* From reference 31

Roberts's overall conclusion is that arms control has a continuing role to play in ensuring international stability. In particular, he points to the importance of creating and maintaining norms of international behavior:

> "...the normative content of arms control is growing increasingly significant. For isolating armed aggressor states, for coping with technology diffusion, and for anchoring US power in common international purposes, the normative aspect of arms control has a special prominence."

In mid-1997 Ambassador James F. Leonard, the US negotiator of the original BTWC, elaborated on how international norms should evolve and considered the question of "The Control of Biological Weapons: Retrospect and Prospect". In his opinion: [32]

> "During the long, slow transition to a new system of international relations, the global and regional arms control treaties of the past 40 years will be complemented by additional treaty regimes covering the full range of conventional weapons..."

Then:

> "...Together, these treaties will provide a framework of norms, obligations, procedures, rules, and interactions that will foster political advances..."

The system will be transformed in the process:

> "...These treaties will bring with them a matrix of verification procedures so penetrating, so ubiquitous, and so intrusive as to be unimaginable today. Total transparency in military matters will be the norm that is steadily and inexorably approached. Military secrecy will be seen, increasingly, as an unhealthy remnant of a previous era in which national security was protected by the balance of power."

If the major driving force which has to be controlled is the biotechnology revolution, then this process in the security field should be reinforced by the growth of environmental [33] and health regimes, [34] as complementary efforts are undertaken in these fields.

5.2 SANCTIONS

Of course, it is not necessarily the case that the international system will evolve in the benign manner envisaged by Ambassador Leonard, in which a positive feedback between regime developments and increased transparency removes the military factor progressively from security concerns. Moreover, even if the general direction of change is positive and benign there are surely likely to be instances in which the international community will be confronted with exceptions to the agreed norm.

There is certainly a wide range of sanctions that could be applied against a violator of the BTWC or the 1925 Geneva Protocol. A potential listing originally suggested to prevent chemical weapons proliferation has been reviewed by Leitenberg. [35] The listing is set out in Table 11. Leitenberg suggests that whilst some of these sanctions are rather trivial, others could have a significant effect on a target country. The range of potential sanctions, however, would allow some matching of the degree of punishment to the severity of the transgression against the norm. [36] The problem, of course, is to find ways of bringing widely agreed and coordinated sanctions to bear effectively on a country.

TABLE 11. Potential sanctions *

- Termination of selected exports or imports
- Complete or selective cutoff of imports and exports
- Loss of access to import preferences
- Loss of civilian export preferences
- Cutoff of bilateral or multilateral economic aid
- A ban on private investment
- A freeze on financial assets
- Expulsion of foreign students in related technical areas
- Refusal to refinance debt
- Termination of airline landing rights
- Embargo on oil imports
- Bans on military sales and assistance
- A reassessment of security ties
- Severance of diplomatic relations
- Loss of political support
- Expulsion from international bodies.

---

* From reference 35

One could hardly find a better illustration of the difficulties faced by the international community than the experience of attempting to use sanctions against Iraq over the last seven years. What is presently known of Iraq's capabilities in regard to biological weapons only began to be understood after the defection of General Husayn Kamil in August 1995. Thereafter: [37]

"...Iraqi officials admitted that they had produced the BW agents anthrax (8,500 liters), botulinum toxin (19,000 liters), and aflatoxin (2,200 liters) after years of claiming that they had conducted only defensive research..."

Leitenberg argues [35] that "Iraq intended to see the sanctions lifted while maintaining its BW program and without complying with the UN resolutions". In the view of the US government, [37] "Saddam's strategy in dealing with UNSCOM is unchanged: he is actively trying to retain what remains of his

WMD programs while wearing down the will of the Security Council to maintain sanctions".

Thus, seven years after the defeat of his armies and the loss of billions of dollars to his economy, sanctions have not worked, although they were authorized by the UN Security Council. Moreover, after the confrontation at the end of 1997, and the deal brokered by the UN Secretary General himself, there has essentially been no change in Iraq's position. UNSCOM concluded after the Vienna Technical Evaluation Meeting (TEM) in March 1998 that: [38]

> "No additional confidence in the veracity and extent of the FFCD was derived from the TEM. Iraq did not provide any new technical information of substance to support its FFCD *[Full, Final and Complete Declaration]*".(Meaning of Acronym added)

It appears that it is the United States that has changed its position in that it is no longer insisting that it will use military force if UN inspectors suffer interference. Dealing with Iraq is reportedly to be left to the Secretary General in order to maintain some semblance of unity in the Security Council. [39]

Thus, whilst sanctions are a potential policy option that needs to be carefully considered, it has to be accepted that only in very exceptional circumstances will they be rapidly effective in correcting a violation.

5.3 ARMED FORCE

It is regularly reported that the armed forces of developed states are expecting to encounter the use of biological weapons in the not too distant future – for example, in a peacekeeping operation. [40] This volume concentrates on the progress being made in the development of detection and protection measures [41] for such forces, despite the undoubted complexity of the technical problems involved. Whilst the continuing uncertainty about Gulf War Syndrome [42] is a cause of great concern, and whilst acknowledging that development of new offensive means is perhaps simpler than providing adequate defense, there must be some hope that progress in defensive measures will reinforce deterrence for some time to come.

But as we know, direct military confrontation on the battlefield, with an identifiable enemy, may be far from the most difficult situation in which a military response to the production or use of biological weapons may be under consideration in the future. The difficulties that the United States has had in convincing its own citizens that it had correctly identified a chemical weapons plant to attack in the Sudan in August of this year, following the terrorist attacks on its embassies in Africa, points to the problems that could be faced by those wishing to use military force in responding to violators.

We have just passed through a period in which advanced, technologically capable states saw weapons of mass destruction as crucial factors in their military arsenals and political power. Now they see the revolution in information technology giving them a massive advantage in conventional conflict. It is hardly surprising, therefore, that they wish to diminish the importance of weapons of mass destruction. But others may have different views. As two US analysts have pointed out, potential opponents have a range of options in responding to such a technological development: [43]

> "...*Emulation* involves replication of forces, typically some sort of mirror-imaging....An *offsetting* response is likely to be a set of countermeasures....*Bypassing* responses involve developing new means of warfare to leapfrog the rival's capabilities, or methods of operation designed to avoid them..." [my emphases]

They go on to suggest that:

> "Weapons of Mass Destruction are a worrisome offset....Powers aspiring to regional hegemony may well conclude such weapons are the best, most credible method available to inhibit great power opposition."

and:

> "Another category of bypassing measures is to conduct warfare in ways not addressable....Guerrilla or terrorist campaigns avoid the force-on-force battles...and designate the American home front as the center of gravity..."

In such circumstances the growing concerns in regard to undeterrable terrorist use of biological weapons are hardly unfounded. [44] Moreover, dealing with the aftermath of a terrorist attack with biological weapons will be very difficult. [45] There could thus be great reluctance to embark on military action. If military action *is* undertaken, it is widely acknowledged that use of nuclear weapons is not, and should not be considered as, a viable option. [36]

6. Conclusion

While other policy options will doubtless be exercised on occasion, the main thrust of our long-term policy should be the achievement of the aims set out by Ambassador Leonard – the increasing regulation of the military aspects of biotechnology. As Geoffrey Vickers has pointed out, achieving such political regulation is a task which will require a very broad consensus based on just those shared values that, at the international level, we so patently lack at present. [46] Yet the enormous benefits that will arise from the application of modern biotechnology, and the opportunities for co-operation – for example, in the control of disease – should give us hope that solutions will be found. [47] Certainly, there does now appear to be consideration of specific practical proposals on the implementation of Article X of the BTWC (on co-operation for economic development) in the ongoing negotiations aimed at strengthening the

280

convention. [48] This should help to ensure more universal adherence to the verification protocol when it is agreed, and that should be the vital next step in the development of this treaty regime.

7. References

1. Spetzel, R.O., Wannemacher, R.W., Sunden, C.D., Franz, D., and Parker, G.W. (1994) *Biological Weapons: Technical Report*, DNA-MIPR-90-715, April, US Army Medical Research Institute of Infectious Diseases, Fort Detrick, MD.
2. Wheelis, M. (1997) Addressing the full range of biological warfare in a BWC Compliance Protocol, paper presented at Pugwash Meeting No. 229, 'Strengthening the Biological Weapons Convention', 20-21 September, Geneva.
3. Miller, D.L. (1952) *History of Air Force Participation in Biological Warfare Program 1944-51*, Historical Study No. 194, September, Historical Office, Office of the Executive, Air Materiel Command, Wright-Patterson Air Force Base, USA.
4. Lebeda, F.J. (1997) Deterrence of biological and chemical warfare: A review of policy options, *Military Medicine* **162**, 156-161.
5. Buchanan, Captain H. L. (1997) Poor man's A-bomb? *Proceedings of the US Naval War College*, April, 83-86.
6. Thränert, O. (1997) Biological weapons and the problem of proliferation, *Aussenpolitik* **11**, 148-157.
7. Pearson, G.S. (1993) Prospects for chemical and biological arms control: The web of deterrence, *The Washington Quarterly* **16 (2)**, 145-162.
8. Dando, M.R. (1997) Technologies for monitoring the Biological and Toxin Weapons Convention: An emerging consensus, paper presented to the NATO ARW on 'Monitoring the Environment for Biological Hazards', 19-22 May, Staszica Palace, Warsaw.
9. Dando, M.R. (1996) The future of biological weapons, paper presented to a NATO ARW on 'The Technology of Biological Arms Control and Disarmament', 28-30 March, Budapest.
10. Dando, M.R. (1996) New developments in biotechnology and their impact on biological warfare, in O. Thränert (ed.), *Enhancing the Biological Weapons Convention*, Dietz, Bonn.
11. Dando, M.R. (1997) Advances in biotechnology: Their relevance to the task of strengthening the Biological and Toxin Weapons Convention, paper presented to a NATO Advanced Studies Institute on 'New Scientific and Technological Aspects of Verification of the Biological and Toxin Weapons Convention', 6-16 July, Budapest.

12. Dando, M.R. (1998) Technological change and future biological weapons, paper presented at a conference on 'Biological Warfare and Disarmament Problems, Perspectives, Possible Solutions', 5-8 July, UNIDIR, Geneva.
13. Dando, M.R. (1999) The impact of the development of modern biology and medicine on the evolution of offensive biological warfare programs in the 20th century, *Defence Analysis* **15**, 43-62.
14. Redmond, C. *et al.* (1998) Deadly relic of the Great War, *Nature* **393**, 747-748.
15. Cohen, W. (1997) *Proliferation: Threat and Response*, Department of Defense, Washington D.C.
    (Text from http://www.defense link.mil/pubs/prolif97/index.html).
16. McEwan, I. (1998) Move over, Darwin... (book review of E. O. Wilson's *Consilience*), *The Observer*, 20 September, p. 20.
17. Ennquez, J. (1998) Policy Forum: Genomics and the world's economy, *Science* **281**, 925-926.
18. Dando, M.R. (1999) The development of international legal constraints on biological warfare in the 20th century, in M. Koskenniemi (ed.), *Finnish Yearbook of International Law*, Martinus Nijhoff, The Hague, pp. 1-69.
19. Editorial (1998) The rules for a biotech revolution, *The Times Higher Education Supplement*, 19 June, p. 11.
20. Kelle, A. (1998) Biotechnology and the development of norms against BW-use, paper presented to a NATO ARW on 'The Role of Biotechnology in Countering BTW Agents', 21-23 October, Prague.
21. Freedman, L. (1992) The end of formal arms control, in E. Adler (ed.), *The International Practice of Arms Control*, Johns Hopkins University Press, Baltimore, pp. 69-84.
22. Sims, J.E. (1992) The American approach to arms control, in E. Adler (ed.), *The International Practice of Arms Control*, Johns Hopkins University Press, Baltimore, pp. 265-287.
23. Rogers, P. and Dando, M.R. (1990) *NBC 90: The Directory of Nuclear, Biological and Chemical Arms and Disarmament 1990*, Tri-Service Press, London.
24. Goldblat, J. (1996) *Arms Control: A Guide to Negotiations and Agreements*, Sage, London.
25. Daalder, I. (1992) The future of arms control, *Survival*, Spring, 51-73.
26. Gallagher, N.W. (1998) Bridging the gaps on arms control, in N.W. Gallagher (ed.), *Arms Control: New Approaches to Theory and Policy*, Frank Cass, London, pp. 1-24.
27. Rattray, G.J. (1996) Introduction, in J.A. Larsen and G.J. Rattray (eds), *Arms Control: Toward the 21st Century*, Lynne Rienner, Boulder, pp. 1-18.
28. Croft, S. (1996) *Strategies of Arms Control: A History and Typology*, Manchester University Press, Manchester.

29. Woodhouse, T., Bruce, R., and Dando, M.R. (1998) *Peacekeeping and Peacemaking: Towards Effective Intervention in Post-Cold War Conflicts*, Macmillan, London.

30. Bertrand, M. (1991) The difficult transformation from 'arms control' into a 'world security system', *International Social Science Journal* **127**, 87-102.

31. Roberts, B. (1997) Arms control in the emerging strategic envionment, *Contemporary Security Policy* **18**, 57-82.

32. Leonard, J. F. (1997) The Control of Biological Weapons: Retrospect and Prospect, Keynote Address at a workshop on 'Inspection Procedures for Compliance Monitoring of the Biological Weapons Convention' (Report ed. J. B. Tucker), Monterey Institute of International Studies, Center for Global Security Research, Lawrence Livermore Laboratory (CGSR-97-002), Livermore, California.

33. Pearson G.S. (1997) The complementary role of environmental and security biological control regimes in the 21st century, *JAMA* **278**, 369-372.

34. Fidler, D.P. (1997) The role of international law in the control of infectious diseases, *Bull. Inst. Pasteur* **95**, 57-72.

35. Leitenberg, M. (1997) Biological weapons, international sanctions and proliferation, *Asian Perspective* **21**, 7-39.

36. Kelle, A. (1998) *Security in a Nuclear Weapons Free World - How to Cope with the Nuclear, Biological and Chemical Weapons Threat*, PRIF Reports No. 50, Peace Research Institute Frankfurt.

37. US Government (1998) *White Paper: Iraqi Weapons of Mass Destruction Programs*, 13 February, Washington D.C.

38. UNSCOM (1998) Report of the United Nations Special Commission's Team to the Technical Evaluation Meeting on the Proscribed Biological Warfare Program (20-27 March 1998, Vienna), United Nations, New York.

39. Erlanger, S. (1998) To rally support, US changes its strategy on Iraq, *International Herald Tribune*, 14 August.

40. Shelton, Col. R.S. (1998) No democracy can feel secure, *Proceedings of the US Naval War College*, August, 39-44.

41. Hewish, M. (1997) Surviving CBW: Detection and protection: What you don't know can kill you, *International Defense Review* **3**, 30-48.

42. Nicholson, G.L. and Nicholson, N.L. (1997) The eight myths of operation 'Desert Storm' and Gulf War Syndrome, *Medicine, Conflict and Survival* **13**, 140-146.

43. Franck, R.E. Jr. and Hildebrandt, G.G. (1996) Competitive aspects of the contemporary military-technical revolution: Potential military rivals to the US, *Defense Analysis* **12**, 239-258.

44. Mann, P. (1998) Officials grapple with 'undeterrable' terrorism, *Aviation Week and Space Technology*, 13 July, 67-70.

45. MacKenzie, D. (1998) Bioarmageddon, *New Scientist*, 19 September, 42-46.

46. Blunden, M. (1995) Vickers and postliteralism, in M. Blunden and M.R. Dando (eds), *Rethinking Public Policy-Making: Questioning Assumptions, Challenging Beliefs*, Sage, London, pp. 11-25.
47. Dando, M.R. (1998) Biotechnology in a peaceful world economy, in E. Geissler, E., Gazsó, L., and Buder, E.(eds), *Conversion of Former BTW Facilities*, Kluwer Academic Publishers, The Netherlands, pp. 25-43.
48. New Zealand and The Netherlands (1999) *Discussion Paper BWC Article X/Protocol Article VII*, Working Paper 362, Ad Hoc Group, United Nations, Geneva, March.

BIOTECHNOLOGY AND THE DEVELOPMENT OF NORMS AGAINST BTW AGENTS

ALEXANDER KELLE
Peace Research Institute Frankfurt
Leimenrode 29
D-60322 Frankfurt/Main

1. Introduction

It may come as a surprise that an edited volume on modern biotechnology and its impact on ways countering BTW agents should conclude with a chapter on "norms", a concept certainly more at home in sociology or political science. However, there are three main areas where modern biotechnology is of particular relevance for the development of norms and rules against BTW agents. The first of these areas is related to the enormous capital invested in commercial biotechnology applications and the huge profits that can be derived from these investments. [1] This might suggest that policy in this field is guided by economic imperatives rather than by normative ones. The second area is related to the "dual use" nature of modern biotechnology, i.e. its possible application in both the civil realm – to make diagnostics, vaccines, medicines – and in the military sphere. [2] This problem is aggravated by the third area, the difficulties involved in distinguishing between perfectly legitimate defensive military R&D and work on an offensive BW program. [3]

To complicate the establishment, maintenance and effectiveness of norms with respect to the way in which biotechnology impacts on efforts to counter BTW agents even further, all these areas share a proclivity towards secrecy. Only if the information related to vaccines or medicines that is not protected by a patent remains secret can the potential profits be realized by industry. Only if the exact nature of BW defensive capabilities remains secret can they be expected to offer protection in case of a BW attack. If they do not, a potential perpetrator attempting to launch a mass casualty attack might use the information available to devise a circumvention strategy. These concerns have led to a very strong emphasis in current efforts to negotiate a Compliance Protocol being placed on the protection of "confidential proprietary information" (CPI), sometimes also called "confidential business information" (CBI) or simply "intellectual property" (IP).

A. Kelle et al. (eds.), The Role of Biotechnology in Countering BTW Agents, 285–297.
© *2001 Kluwer Academic Publishers. Printed in the Netherlands.*

It will be argued in this chapter that the way in which biotech and pharmaceutical industries insist on the protection of their CPI / IP, together with the receptiveness of some governments to being made to serve the interests of their respective industries, threatens to lead to a Protocol which does not strengthen or develop further norms against BTW agents, but might instead lead to their weakening and undermining.

In addition, and related to the second and third characteristic mentioned above, the chapter will address the dual-use problem and the offense – defense dichotomy as they relate to the establishment and setting up of normative structures which can guide behavior in efforts to counter BTW agents. Here, the emphasis will be on what individuals or groups involved in these two areas can do to act in line with, complement, and strengthen the norms of the regime.

However, before discussing the impact of biotechnology on the development of norms in the context of the BTW control regime, I will outline the meaning of the term "norm" as it will be used throughout this chapter, place norms in the wider context of regime analysis, and present the structure of the "old" BTW control regime without the compliance protocol currently being negotiated.

2. Norms in International Regimes

International regimes are cooperative institutions with a largely formal structure which is either legally codified or not. Conceptualizing international regimes as institutions avoids confusing them with either international organizations or international treaties. Yet, many regimes have their structure formalized in a treaty and utilize an international organization in order to put the regime into effect and verify compliance by regime members, which are almost exclusively states. [4]

According to a definition which is used as a point of reference by many studies of international regimes, norms are one of four distinct structural elements of an international regime. Krasner summarizes these as follows:

> "Principles are beliefs of fact, causation, and rectitude. **Norms** are standards of behavior defined in terms of rights and obligations. Rules are specific prescriptions or proscriptions for actions. ... procedures are prevailing practices for making and implementing collective choice." [5, p. 1] (emphasis added)

While principles and norms account for the basic structure of a regime, quite diverse rules and procedures can be envisioned to put into effect the regime's principles and norms. In more practical terms, "fundamental political arguments are more concerned with norms and principles than with rules and procedures." [5] However, it is quite possible that disputes about the concretization of norms through rules and procedures only mask more fundamental disagreements over the utility and importance of norms or even principles of an international regime.

In addition to debates among regime members about the proper functioning of the regime, one can expect discourses within member states about regime adherence. These discourses occur almost of necessity because regime-guided behavior amounts to nothing less than abandoning policy choices based upon the narrow calculation of self-interest. Instead, regime-compliant behavior by states places a higher value upon cooperatively agreed upon norms as yardsticks for action. Especially in the field of national security policy, this is not an easy choice to make. Therefore the "internalization" of the principles, norms, and rules of the regime by the states is an important prerequisite for regime-compliant behavior. Müller identifies four "general-normative reflexive nesting complexes" which largely influence the internalization of international regimes and on which regime proponents in domestic debates can draw: first, a general hierarchical framework of international norms which are not issue specific; second, international law with the norm *pacta sunt servanda*; third, status and prestige that is conferred by taking over responsibility for the regime; fourth, domestic law which translates international treaty commitments into the domestic realm. [6]

3.   The Structure of the BTW Control Regime

3.1 PRINCIPLES UNDERLYING THE BTW CONTROL REGIME

If one looks first at the principles underlying the BTW control regime, four can easily be identified with a fifth possibly waiting in the wings of the compliance protocol currently being negotiated. The first principle is related to the fact that the use of BTW agents constitutes an abhorrent act of warfare. Most commonly, this principle is referred to as the "BW taboo". The second principle on which the BTW control regime is based defines peaceful uses of the biosciences as a legitimate undertaking. The third principle expresses the assumption of states subscribing to the regime that defense against the threat or use of BW is permitted. This principle is rooted in the belief that the peaceful uses of biosciences cannot be taken for granted. However, while this underlying belief has led in both the nuclear non-proliferation regime and the CW control regime to the manifestation of yet another principle, the verification principle, this has not up to now materialized in the BW context. Neither the Geneva Protocol nor the BWC makes explicit reference to such a principle. Even if one takes a closer look at the ongoing negotiations for a compliance protocol, so far it appears less than certain that a verification principle is accepted by all negotiating parties (see below). Last but not least, the fourth principle underlying the BWC control regime is the complementarity principle which is clearly stated in preambular paragraphs 2 to 4 of the BWC: the BWC complements the 1925 Geneva Protocol. Nothing in the former can be construed to contradict the content of the latter.

3.2 NORMS OF THE BTW CONTROL REGIME

In addition to the already mentioned *non-use norm,* which is central to the BW control regime, the *non-acquisition norm* - in contrast to the nuclear nonproliferation regime - applies *erga omnes*, as does the exception of activities for defensive purposes. The corresponding provision is to be found in Article I of the BW convention. The *non-transfer norm* is contained in Article III of the BWC. It is furthermore put into effect through export control measures by individual states and - since 1990 - through the activities of the Australia Group. The *disarmament norm* is clearly spelled out in Article II of the BWC, which requires that all states parties either destroy or divert to peaceful purposes all agents, toxins, equipment and means of delivery related to their BW holdings.

In addition, the *cooperation norm* is spelled out in Article X of the BWC, the *assistance norm* is contained in Article VII, and the normative requirement to *continue negotiating a CW treaty* was written into BWC Article IX.

The *consultation norm* is spelled out in Article V of the BWC, in which states parties agree to

> "consult one another and to co-operate in solving any problems which may arise in relation to the objective of, or in the application of the provisions of, the Convention." [Quoted from reference 7, p. 21]

What is more, the *consultation norm*, albeit in a somewhat different form, also guides the behavior of states participating in the Australia Group, which have agreed to consult one another in case of export denials of certain dual-use items and technologies to states of proliferation concern. [2]

Lastly, the *inspection norm* is practically absent from the existing BW control regime. Justified as this omission may have been at the time of the conclusion of the BWC, the clandestine offensive BW programs of Iraq and the Soviet Union have clearly shown that the hypothesis of military disinterest in BW no longer correspond to a reality in which even a BWC depository engaged in prohibited activities on a massive scale. This absence of both the inspection norm and of any concrete rules which could put the norm into effect in everyday state practice is all the more disturbing as confidence in the implementation of other regime norms also depends on some form of inspection.

3.3 THE MISSING PARTS OF THE CURRENT REGIME: RULES AND PROCEDURES

This points to the missing parts of the BTW control regime, i.e. the rules and procedures for putting into effect the more abstract principles and norms. It also identifies the area in which the Ad Hoc Group (AHG) can make the biggest contribution to the strengthening and further development of the regime structures. However, one has to avoid the pitfall of assuming that any rules spelling out more concretely the less specific guidance contained in the regime norms are good rules. This certainly is not the case. Rather, rules and norms have to be related in such a way that the former do not leave the latter as empty shells or, even worse, contradict them.

As to the few existing rules and procedures of the BTW control regime that can be identified, they are exclusively related to the organization and governance of the BWC. Article XI contains the procedures for amending the Convention, Article XII spells out the initial review procedure, according to which five years after its entry into force a conference of states parties to the BWC shall be held. Furthermore, the withdrawal procedure is set out in Article XIII, 2 and Article XIV contains the stipulations for ratification of, accession to and entry into force of the BWC.

4. Modern Biotechnology, Security Interests and Norms Against BTW Agents

Modern biotechnology will – at least to some extent – be subjected to the compliance measures negotiated in the Ad Hoc Group. When this was recognized by biotechnology and pharmaceutical industries, their respective industry associations started to formulate their own positions which were then brought to the attention of government authorities and the negotiators in Geneva.

The U.S. industry was the first to go public with its position paper issued during the Fourth Review Conference of the BWC at the end of 1996. [8] Because of this leading position and due to its size and the fact that the U.S. biotech sector is the one on which most information is available, this chapter will focus on the positions taken by U.S. biotech industry and how these positions relate to the further development of the norms guiding state action in the BW control regime. [9, 10]

As regards the non-use norm, there is certainly no objection being made by any biotech industry to this stipulation. The same can be said for the non-acquisition, the non-transfer and the disarmament norms. Along these lines a U.S. industry representative declared that industry is "strongly supportive of the intent of the BWC", accepts the need to "develop a practical compliance protocol as an adjunct document to the original BWC treaty", and supports "most of the moves that serve to reduce the threat posed by biological weapons." [11] In principle, this support is extended to measures directed at putting into effect an inspection norm as well. However, the details of the inspection rules and procedures which industry is willing to accept cannot but raise the question of the extent to which the proposed measures are suitable for actually developing a strong and robust inspection norm.

In its November 1996 statement before the Fourth Review Conference of the BWC, the U.S. Pharmaceutical Research and Manufacturers Association (PhRMA) argued for

- making declarations under the 1986 and 1991 Confidence Building Measures mandatory, while limiting their scope to the already agreed upon measures; [12]
- a limitation of on-site measures to challenge inspections only. Non-challenge visits were rejected; [13]
- subjecting a request for a challenge visit to a "green-light" procedure in which a majority of three quarters of the members of the executive body of the future BWC organization would have to approve any request.

Although some details of the PhRMA position have been changed since then, the principles underlying the approach taken by the industry association have not.

This is problematic in so far as, as one scholar has pointed out, "the measures proposed by PhRMA, if implemented in a Protocol to the BWC, would be very unlikely to provide adequate assurance that parties were living up to their obligations." [9] However, PhRMA's approach is not only informed by generic concerns about CPI / IP but also by its experiences with inspections that were carried out in the context of the Trilateral Process – agreed among the U.S., the United Kingdom, and Russia – in two U.S. industry facilities. After both visits members of the Russian team claimed that offensive BW research at the two sites could not be excluded. Thus, according to one analyst

> "[f]or the US pharmaceutical industry, these trilateral inspections constituted an ominous introduction to what efforts to monitor compliance with the BWC might be like. ... To compound the situation, a negative atmosphere seeped into the interactions between some US government and industry officials. ... From Industry's perspective, the government's failure to conclude a negotiating position was a sign of more trouble ahead." [9]

In response to this domestic environment, the U.S. pharmaceutical industry started in 1997 to lobby its foreign counterparts so as to raise their awareness of the issue. This effort to harmonize positions culminated in a May 2000 joint statement by European, Japanese and U.S. pharmaceutical, chemical and biotechnological industries. In this joint paper it is claimed that "any routine on-site activity is not a useful concept under the Protocol." [8]

When the U.S. government eventually issued its position paper in January 1998 in the form of a White House fact sheet, it showed a considerable overlap with the positions taken by U.S. industry. The U.S. initiative contained four elements: declarations "about facilities and/or activities that are especially suited for possible BW purposes"; voluntary visits, which can take place at the invitation of the visited facility; non-challenge clarifying visits aiming at clarification of "an ambiguity, uncertainty, anomaly, omission or other issue related to" member states annual declarations; and challenge investigations that "should be subject to a 'green light' filter, under which a simple majority of the governing body" has to approve an investigation request. [14]

It is not very likely that such an arrangement, containing only focussed non-challenge visits, would provide all the benefits one could expect from a regime which would include random non-challenge visits as well. At least, the deterrence value of an inspection system consisting of focussed non-challenge visits (NCV) only has to be strongly questioned. Furthermore, the efficiency, preparedness and professional standing of the future BTWC organization's inspectorate can be expected to suffer from such a set of inspection rules. [13] In sum, the first area in which biotechnology impacts on norm development and strengthening clearly shows a trend towards a minimalist transfer of authority to the future BTWC organization and the retention of as much freedom of action for national measures as possible.

This trend away from cooperative norm-based measures towards unilateral action continues in the context of protecting national security interests as well. Again, the U.S. seems to be in the lead when it comes to placing less emphasis on a strong regulatory arms control framework. The U.S. government is pursuing and investing heavily in a unilaterally oriented policy that favors deterrence and defense over arms control. Measures that bear witness to this policy include biodefense measures for U.S. troops abroad and the U.S. population at home, and the opaqueness introduced into U.S. nuclear strategy when it comes to the retaliatory use of nuclear weapons against BW first use. [15] In addition, the establishment of a national missile defense system is being justified by some of its proponents with the threat from ballistic missiles armed with CBW warheads. This trend towards unilateralism seems to grow stronger when the primary threat scenario involving BW is that of a terrorist attack. [16] Against bioterrorism, however, conventional wisdom argues that multilaterally agreed upon arms control norms and rules are of little to no value. Instead, defending the home front is the primary task. [17]

When observing the AHG negotiations one cannot avoid the impression that some strange bedfellows, i.e. the United States and some of the more radical states of the Non-Aligned Movement (NAM), have come together with the common goal of allowing only weak non-challenge on-site measures into the compliance protocol. From the viewpoint of regime analysis this works hand in hand with industry interests in leaving the normative structure of the BTW control regime with only a weak regulatory (rule-based) underpinning. [18]

5. Sources of Individual or Group Action Against the Misuse of the Biosciences

In addition to the international and state levels of norm-guided behavior against BW agents, individual and/or group action against the misuse of the biosciences in general and modern biotechnology in particular have to be considered. Three sources for such norm-guided behavior of groups or individuals can be distinguished: first, the already mentioned internalization of regime norms; second, bioethics; third, the criminalization of biological warfare.

5.1 INTERNALIZATION OF REGIME NORMS

Of the four "reflexive nesting complexes" identified by Müller as crucial for the internalization of regime norms, the first one – a hierarchical complex of norms on the macro, i.e. the international, level can hardly be expected to serve as a guide for action on the micro-level of individual or group action against BTW agents. The same can be assumed to hold for status and prestige. A somewhat stronger source for norm-guided behavior can be found in international law, the second regime nesting complex. Although primarily aimed at conditioning state behavior, it may well influence the actions of sub-national actors. As Colwell and Zilinskas have pointed out, the BWC formulates an ethical standard not to violate its provisions, even for citizens of non-state parties. [19] However, the strongest normative frame of reference for regime-compliant behavior can be expected to emanate from domestic law. It is therefore of crucial importance that the BWC protocol not only contain provisions which make it obligatory for states parties to enact domestic legislation, but also to have the future BTWC organisation check on compliance with such requirements. Assistance in drawing up domestic legislation might be provided by the secretariat of the future organization for states parties who request this.

5.2 BIOETHICS AND THE MILITARY USE OF THE BIOSCIENCES

Scientists who are morally aware should certainly be concerned with possible applications of their research and devise their research programs in such a way as to further the well-being of and not to harm humans. [20] Yet, with respect to biotechnology and biological warfare, things are not that straightforward. The offense-defense dilemma has already been referred to elsewhere in this volume, and things are complicated further by the fact that

> "under some circumstances, research cannot be differentiated from development. For example, if an outside analyst were to inspect a facility that developed a virus vaccine, when observing research activities that analyst would be faced with two difficult problems: (1) determining if research was aimed at developing an attenuated agent for a vaccine or a more virulent agent for BW and (2) ascertaining when the research phase ceases and the development phase commences." [17]

It is because of this loophole, i.e. the omission of research from the list of prohibited activities in the BWC, that Lappé has strongly advocated a moral analysis of the acceptability of BW research, of which he distinguishes three categories: first, research in direct support of offensive BW weapon systems; second, defensive research that leads to both a defensive and an offensive capability; third, "purely" defensive research. [3] The first category is obviously morally unacceptable and for citizens of BWC states parties legally prohibited. Research of both defensive and offensive utility can be considered ethical only in cases in which defensive applications clearly outweigh the offensive ones. With respect to defensive research, six areas on which moral judgement should be passed can be identified. They are:

"(i) isolation of preparations of toxins and pathogenic organisms;
(ii) propylactic measures (vaccines, toxoids);
(iii) therapeutic measures (antibiotics, antiviral drugs, antidotes)
(iv) protective clothing and equipment;
(v) monitoring and detection devices;
(vi) methods of decontamination." [3, p. 82]

In sum, judgement should be based on four indicators: first, the likely proximity of research to weapons development and production. This applies to defensive technologies as well, since some of these could act as enabling factors for offensive military activities. Second, the mission of the sponsoring agency has to be taken into account. Third, it has to be considered how great the contribution of the research to peaceful and beneficial ends is. The permission to publish the research results freely serves as a fourth yardstick. [3]

Yet, despite the formulation of such standards, offensive BW programs have been initiated and in all likelihood are still continuing in some parts of the world. This leads first to the question of what motivates scientists to promote and take part in such activities. After that the question of how such participation can be prevented has to be addressed. As regards the first question, Colwell and Zilinskas identify six reasons for scientists to participate in offensive BW-programs: already existing scientific interest in infectious disease, the scientific challenge to make BW actually work, personal enrichment, job security, making a contribution to national security, and threats made against scientists or their relatives.

The last two motivational factors are best dealt with by addressing the incentive structure of the state considering BW acquisition through a set of measures usually summed up under the heading "demand side" non-proliferation policy. Such a policy includes *inter alia* addressing the security concerns of the potential proliferant states, offering civilian incentives as compensation for military restraint, and the like. To the extent that such measures succeed in reducing the incentive for a state to acquire BW, that state can be expected to refrain from exerting pressure on its scientists to participate in an offensive BW program. Similarly, if national security is not defined in terms of the possession of BW, scientists will find it difficult to identify themselves with such a policy.

In order to address the problem of job security, Colwell and Zilinskas suggest that scientists with expertise in "specialized fields such as endo- or exotoxins, specific arboviral strains, or aerobiology [...] become involved in the work of national or international agencies responsible for controlling biological weapons proliferation and detecting BW programs." [17] While this proposal certainly deserves support, one has to acknowledge at the same time that the capacity of such agencies to integrate scientists with relevant expertise is limited. Even if the jobs of additional scientists can be secured through conversion efforts [21], it is obvious that not all researchers involved in an offensive BW program of, for example, the size of that of the Soviet Union can be accommodated through these measures.

This is where ethical considerations enter the picture again. The crucial task is then to communicate the moral standards outlined above to scientists who might be tempted to work in a clandestine offensive BW program. One of the most efficient institutions during the Cold War, "able to bridge ideological gaps in bringing together scientists from different political and social systems in neutral settings" was the Pugwash Conference on Science and World Affairs. With a long tradition of organizing BW-related workshops [22], Pugwash can be expected to continue as one of the most efficient "transmission belts" for conveying ethical standards in the future.

5.3 CRIMINALIZATION OF BIOLOGICAL WARFARE

An additional normative signpost for individuals' actions could be established through the criminalization of BW activities. An international "Convention to Prohibit Biological and Chemical Weapons under International Criminal Law" has been proposed by the Harvard Sussex Program on CBW Armament and Arms Limitation. [23, 24] The draft convention would build upon and expand provisions contained in the CWC (Article VII) and BWC (Article IV). Both BWC and CWC require – to differing degrees – that states parties incorporate the prohibitions contained in them into domestic law.

> "However the BWC and CWC do not attempt to make the development, production, possession or use of biological and chemical weapons an international crime for which states establish jurisdiction over prohibited acts regardless of the place where they are committed or the nationality of the offender, nor do these treaties contain provisions dealing with the extradition of suspects." [21]

A convention to prohibit CBW under international criminal law would place the above mentioned CBW-related activities in the same category as aircraft hijacking, hostage taking, torture, and theft of nuclear materials, to mention just a few especially repugnant crimes for which the international community has already concluded comparable conventions.

The convention would use the definitions of BWC and CWC, thereby reiterating the general purpose criterion, and its scope would cover all activities prohibited in these two conventions even if conducted by an official or employee of a state not party to either BWC or CWC or the proposed convention on criminalization. According to its authors:

"Adoption and widespread adherence to such a convention would create a new dimension of constraint against biological and chemical weapons by applying international criminal law to hold individual offenders responsible and punishable wherever they may be and regardless of whether they act under or outside of state authority." [21]

6. Conclusion

Current efforts to strengthen the BWC through the conclusion of a Compliance Protocol can be expected to lag behind the most effective conceivable normative and regulatory framework for combating the misuse of the biosciences in general and biotechnology in particular. Especially the weak provisions for non-challenge visits are likely to create a regulatory loophole in the protocol.

In case this approach prevails in the negotiations, we will very likely see a BWC Compliance Protocol which contains an inspection norm, which by itself is a considerable step forward that should not be belittled. However, the rules and procedure that are being devised to put this norm into effect are unlikely to give the inspection norm the teeth it needs in order not to appear as a paper tiger. This can be expected to lead to a strong reliance on unilateral action, i.e. intelligence gathering, defense and deterrence policies, at the expense of cooperative efforts. Since such a course of action can only be pursued by a small number of BWC states parties, frictions among the convention's membership would not be surprising.

Such frictions and conflicts may well spill over into the realm of moral and ethical norms or the criminalization of offensive BW activities under international criminal law as signposts for individual or group action. It is difficult to imagine that negotiating an international convention to criminalize offensive BW activities will find many state supporters if the BWC and its compliance protocol are a source of friction and conflict. However, finding some countries that push such negotiations is deemed essential by the authors of the existing draft convention. [21] Likewise, if the BWC continues to be perceived and/or portrayed by some states as discriminating against them, these states will certainly find ways of preventing their scientists from following universal ethical standards which aim at applying biosciences in order to benefit rather than harm mankind.

This would be all the more troublesome as these two sources would be urgently needed as additional normative constraints on the offensive military application of the biosciences, should the weak normative frame of reference currently being negotiated in the Ad Hoc Group actually materialize.

296

7. References

1. Holmberg, A. (1994) Industry Concerns Regarding Disclosure of Proprietary Information, in *Director Series on Proliferation*, No. 4, Lawrence Livermore National Laboratory, Livermore, pp. 91-99.
2. Robinson, J.P. (1996) Some Lessons for the Biological Weapons Convention from Preparations to Implement the Chemical Weapons Convention. in O. Thränert (ed.), *Enhancing the Biological Weapons Convention*, Verlag J.H.W. Dietz Nachfolger, Bonn, pp. 86-113.
3. Lappé, M. (1990) Ethics in Biological Warfare Research, in S. Wright (ed.), *Preventing A Biological Arms Race,*: The MIT Press, Cambridge, pp. 78-99.
4. Müller, H. (1993) *Die Chance der Kooperation. Regime in den Internationalen Beziehungen*, Wissenschaftliche Buchgesellschaft, Darmstadt.
5. Krasner, S. (1983) Structural Causes and Regime Consequences: Regimes as Intervening Variables, in S. Krasner (ed.), *International Regimes*, Cornell University Press, Ithaca, pp. 1-25.
6. Müller, H. (1993) The Internalization of Principles, Norms, and Rules by Governments. The Case of Security Regimes, in V. Rittberger with P. Mayer (ed.), *Regime Theory and International Relations*, Clarendon Press, Oxford, pp. 361-388.
7. Sims, N. (1988) *The Diplomacy of Biological Disarmament. Vicissitudes of a Treaty in Force, 1975-1985*, St. Martin's Press, New York.
8. PhRMA-website at http://www.phrma.org/srpub/bwc.html#phrma.
9. Smithson, A.E. (1998) Man versus Microbe: The Negotiations to Strengthen the Biological Weapons Convention, in *Biological Weapons Proliferation: Reasons for Concern, Courses of Action*, Report No. 24, The Henry L. Stimson Center, Washington, D.C., pp. 107-128.
10. Dando, M. (1998) *Implications of a strengthened biological and toxin weapons convention for the biotechnology and pharmaceutical industry*, paper prepared for the Conference "A Strengthened BTWC - Potential Implications for Biotechnology", 28 - 29 May, Vienna, Austria.
11. Muth, W.L. (1999) The Role of the Pharmaceutical and Biotech Industries in Strengthening the Biological Disarmament Regime, *Politics and the Life Sciences* **18 (1)**, 92-97.
12. Hunger, I. (1996) Confidence Building Measures, in G. Pearson/M. Dando (eds.), *Briefing Book for the Fourth Review Conference of the BWC*, Quaker United Nations Office, Geneva.
13. Pearson, G.S. (1997) *Strengthening the Biological Weapons Convention: The Necessity for Non-Challenge Visits*, Briefing Paper No. 2, University of Bradford, Bradford.
14. The White House, Office of the Press Secretary (1998) *Fact Sheet: The Biological Weapons Convention*, Washington, D.C., 27 January.
15. Thränert, O. (1998) *Zwischen Rüstungskontrolle und Abschreckung: Ame-*

rikanische Antworten auf die Verbreitung Biologischer Waffen, Studie zur Außenpolitik, No.73, Friedrich-Ebert-Stiftung, Bonn.

16. Roberts, B. (ed.) (2000*) Hype or Reality? The New Terrorism and Mass Casualty Attacks*, The Chemical and Biological Arms Control Institute (CBACI), Alexandria, VA.

17. Lugar, R. (1998) *Countering Chemical and Biological Weapons: The Need for a Defense in Depth*, The Arena, No. 7, CBACI, Alexandria, VA.

18. Wilson, H. (1999) BWC Update, *Disarmament Diplomacy* **42**, 27-34.

19. Colwell R.R. and Zilinskas, R.A. (2000) Bioethics and the Prevention of Biological Warfare, in R.A. Zilinskas (ed.), *Biological Warfare - Modern Offense and Defense*, Lynne Rienner, Boulder, pp. 225-245.

20. Sinsheimer, R.L. (1990) Scientists and Research, in S. Wright (ed.), *Preventing a Biological Arms Race,*: The MIT Press, Cambridge, pp. 71-77.

21. Buder, E., Geissler, E., and Gazsó, L. (eds.) (1998*) Conversion of Former BTW Facilities*, NATO-Science Series, Kluwer Academic Publishers, Dordrecht.

22. Robinson, J.P.P. (1998) The Impact of Pugwash on the Debates over Chemical and Biological Weapons, in *Scientific Cooperation, State Conflict: The Roles of Scientists in Mitigating International Discord*, Vol. 866, Annals of the New York Academy of Sciences, 224-252.

23. A Draft Convention to Prohibit Biological and Chemical Weapons under International Criminal Law, Editorial (1998) *The Chemical and Biological Weapons Conventions Bulletin* **42**, 1-2.

24. Draft Convention on the Prevention and Punishment of the Crime of Developing, Producing, Acquiring, Stockpiling, Retaining, Transferring or Using Biological or Chemical Weapons (1998), *The CBW Conventions Bulletin* **42**, 1998, 2-5.

Index

ABM Treaty 274
acetylcholine receptor 106
acidification rate 117
active defense(s) 23, 37-38
Ad Hoc Group (AHG) 9, 13, 16-
  19, 21, 27-29, 38-43, 49, 289,
  291, 295
adjuvant/adjuvants 108, 165, 167-
  169, 172, 174-175, 181, 187, 193
aerobiology 53, 271, 294
aerosol/aerosols 38, 168, 185, 211-
  212, 220, 222, 228, 239, 247
affinity selection 111
aflatoxin 153, 278
aggressor 37, 173, 197-198, 200,
  212, 233, 268-269, 276
agriculture 48, 54, 106
AHG *see Ad Hoc Group*
amantadine 252
amperometric 110
amplification 84-86, 93-95, 97-
  100, 112-114, 208, 218-219
analysis/analyses and identification
  2, 3-4, 6
  off-site ~ 145-147, 155
  on-site ~ 100-101, 147, 154-156
analytical laboratory techniques
  108
anaphylactic activity 248
anaphylactic shock 251
animal immune plasma 248
animal viruses 220
animals 9, 65, 71, 106, 108, 110-
  111, 144, 152, 154, 186, 188, 199,
  207, 209, 215, 217, 219, 248-249
annealing 82, 94, 112
Annex on Chemicals to the CWC
  39
anthrax 10-11, 38, 43, 72, 107-108,
  153, 171, 185-187, 200, 233, 270,
  277
anthropozoonotic 58
antibiogram 123
antibiotic/antibiotics 10, 121-122,

131, 171, 179, 181, 186, 200, 208,
  230, 232, 238-240, 271, 293
antibiotic/immunotherapy 124
antibody/antibodies 69-73, 75-76,
  85, 96-98, 106, 110-111, 113-114,
  117, 123-124, 126-128, 162-163,
  165-166, 169, 171-172, 186, 189,
  192-193, 195, 231, 248-251
antibody-based assay 69
antidotes 38, 272
antigen 69, 71-72, 73, 75-76, 96-
  97, 106, 111, 114, 117, 126, 136,
  162-165, 167-169, 171-173, 176-
  178, 181, 187, 189, 190, 195, 197-
  201, 229, 231
antisera 211, 231
arboviral strains 294
arboviruses 58
Arenaviruses 220-221
arms control 1-2, 4-5, 9, 33, 222,
  270, 273-276, 291
arthralgia 170, 250
Article X 16-17, 21, 30, 288-289
assassination 269
atomic force microscopy (AFM) 76
Aum Shinrikyo 13, 28, 236
Australia Group 22, 288
automation 89, 100, 101, 133, 136
automatons 125
autoradiography 84

B lymphocyte 111, 162
baboons 250-251
Bacillus anthracis 72, 123, 125,
  130, 165, 171, 185-187, 191, 194-
  195, 239
bacteria/bacterium 10, 69-70, 76,
  81, 84, 85-87, 94, 99, 107, 109,
  110-112, 122-125, 152, 161, 164-
  165, 167-168, 170-171, 187, 189,
  195, 197, 200, 210, 212, 228-231,
  238, 240, 247, 250
bacteriophage 93, 110
ballistic missiles 37-38, 228, 291

9 780792 369066